MINI MOKE
Gold Portfolio
1964 -1994

Compiled by
Tim Nuttall of the Mini Moke Club

ISBN 1 85520 2409

BROOKLANDS BOOKS LTD.
P.O. BOX 146, COBHAM,
SURREY, KT11 1LG. UK

Printed in Hong Kong

BROOKLANDS BOOKS

BROOKLANDS ROAD TEST SERIES

Abarth Gold Portfolio 1950-1971
AC Ace & Aceca 1953-1983
Alfa Romeo Giulietta Gold Portfolio 1954-1965
Alfa Romeo Giulia Berlinas 1962-1976
Alfa Romeo Giulia Coupés 1963-1976
Alfa Romeo Giulia Coupés Gold P. 1963-1976
Alfa Romeo Spider 1966-1990
Alfa Romeo Spider Gold Portfolio 1966-1991
Alfa Romeo Alfasud 1972-1984
Alfa Romeo Alfetta Gold Portfolio 1972-1987
Alfa Romeo Alfetta GTV6 1980-1987
Allard Gold Portfolio 1937-1959
Alvis Gold Portfolio 1919-1967
Armstrong Siddeley Gold Portfolio 1945-1960
Aston Martin Gold Portfolio 1972-1985
Austin Seven 1922-1982
Austin A30 & A35 1951-1962
Austin Healey 100 & 100/6 Gold P. 1952-1959
Austin Healey 3000 Gold Portfolio 1959-1967
Austin Healey Sprite 1958-1971
BMW Six Cyl. Coupés 1969-1975
BMW 1600 Collection No.1 1966-1981
BMW 2002 Gold Portfolio1968-1976
BMW 316, 318, 320 (4 cyl.) Gold P. 1975-1990
BMW 320, 323, 325 (6 cyl.) Gold P .1977-1990
BMW 5 Series Gold Portfolio1981-1987
BMW M Series Performance Portfolio1976-1993
Bristol Cars Gold Portfolio 1946-1992
Buick Automobiles 1947-1960
Buick Muscle Cars 1965-1970
Cadillac Automobiles 1949-1959
Cadillac Automobiles 1960-1969
Chevrolet 1955-1957
Chevrolet Impala & SS 1958-1971
Chevrolet Corvair 1959-1969
Chevy El Camino & SS 1959-1987
Chevy II Nova & SS 1962-1973
Chevelle & SS Muscle Portfolio 1964-1972
Chevrolet Muscle Cars 1966-1971
Chevy Blazer 1969-1981
Chevrolet Corvette Gold Portfolio 1953-1962
Chevrolet Corvette Sting Ray Gold P. 1963-1967
Chevrolet Corvette Gold Portfolio 1968-1977
High Performance Corvettes 1983-1989
Camaro Muscle Portfolio 1967-1973
Chevrolet Camaro Z28 & SS 1966-1973
Chevrolet Camaro & Z28 1973-1981
High Performance Camaros 1982-1988
Chrysler 300 Gold Portfolio 1955-1970
Chrysler Valiant 1960-1962
Citroen Traction Avant Gold Portfolio 1934-1957
Citroen 2CV Gold Portfolio 1948-1989
Citroen DS & ID 1955-1975
Citroen DS & ID Gold Portfolio 1955-1975
Citroen SM 1970-1975
Cobras & Replicas 1962-1983
Shelby Cobra Gold Portfolio 1962-1969
Cobras & Cobra Replicas Gold P. 1962-1989
Cunningham Automobiles 1951-1955
Daimler SP250 Sports & V-8 250 Saloon Gold Portfolio 1959-1969
Datsun Roadsters 1962-1971
Datsun 240Z 1970-1973
Datsun 280Z & ZX 1975-1983
The De Lorean 1977-1993
De Tomaso Collection No. 1 1962-1981
Dodge Charger 1966-1974
Dodge Muscle Cars 1967-1970
Dodge Viper on the Road
Excalibur Collection No. 1 1952-1981
Facel Vega 1954-1964
Ferrari Cars 1946-1956
Ferrari Collection No. 1 1960-1970
Ferrari Dino 1965-1974
Ferrari Dino 308 1974-1979
Ferrari 308 & Mondial 1980-1984
Motor & T&CC Ferrari 1966-1976
Motor & T&CC Ferrari 1976-1984
Fiat 500 Gold Portfolio 1936-1972
Fiat Pininfarina 124 & 2000 Spider 1968-1985
Fiat-Bertone X1/9 1973-1988
Ford Consul, Zephyr, Zodiac Mk.I & II 1950-1962
Ford Zephyr, Zodiac, Executive, Mk.III & Mk.IV 1962-1971
Ford Cortina 1600E & GT 1967-1970
High Performance Capris Gold P. 1969-1987
Capri Muscle Portfolio 1974-1987
High Performance Fiestas 1979-1991
High Performance Escorts Mk.I 1968-1974
High Performance Escorts Mk.II 1975-1980
High Performance Escorts 1980-1985
High Performance Escorts 1985-1990
High Performance Sierras & Merkurs Gold Portfolio 1983-1990
Ford Automobiles 1949-1959
Ford Fairlane 1955-1970
Ford Ranchero 1957-1959
Thunderbird 1955-1957
Thunderbird 1958-1963
Thunderbird 1964-1976
Ford Falcon 1960-1970
Ford GT40 Gold Portfolio 1964-1987
Ford Bronco 1966-1977
Ford Bronco 1978-1988
Holden 1948-1962
Honda CRX 1983-1987
Hudson & Railton 1936-1940
Isetta 1953-1964
ISO & Bizzarrini Gold Portfolio 1962-1974
Jaguar and SS Gold Portfolio 1931-1951
Jaguar XK120, 140, 150 Gold P. 1948-1960
Jaguar Mk.VII, VIII, IX, X, 420 Gold P.1950-1970
Jaguar 1957-1961
Jaguar Mk.2 1959-1969
Jaguar Cars 1961-1964
Jaguar E-Type Gold Portfolio 1961-1971
Jaguar E-Type 1966-1971
Jaguar E-Type V-12 1971-1975
Jaguar XJ12, XJ5.3, V12 Gold P. 1972-1990
Jaguar XJ6 Series II 1973-1979
Jaguar XJ6 Series III 1979-1986
Jaguar XJS Gold Portfolio 1975-1990
Jeep CJ5 & CJ6 1960-1976
Jeep CJ5 & CJ7 1976-1986
Jensen Cars 1946-1967
Jensen Cars 1967-1979
Jensen Interceptor Gold Portfolio 1966-1986
Jensen Healey 1972-1976
Lagonda Gold Portfolio 1919-1964
Lamborghini Cars 1964-1970
Lamborghini Countach & Urraco 1974-1980
Lamborghini Countach & Jalpa 1980-1985
Lancia Beta Gold Portfolio 1972-1984
Lancia Fulvia Gold Portfolio 1963-1976
Lancia Stratos 1972-1985
Land Rover Series I 1948-1958
Land Rover Series II & IIa 1958-1971
Land Rover Series III 1971-1985
Land Rover 90 & 110 1983-1989
Lincoln Gold Portfolio 1949-1960
Lincoln Continental 1961-1969
Lincoln Continental 1969-1976
Lotus & Caterham Seven Gold P. 1957-1993
Lotus Sports Racers Gold Portfolio 1953-1965
Lotus Elite 1957-1964
Lotus Elite & Eclat 1974-1982
Lotus Elan Gold Portfolio 1962-1974
Lotus Elan Collection No. 2 1963-1972
Lotus Cortina Gold Portfolio 1963-1970
Lotus Europa Gold Portfolio 1966-1975
Lotus Turbo Esprit 1980-1986
Motor & T&CC on Lotus 1979-1983
Marcos Cars 1960-1988
Maserati 1965-1970
Maserati 1970-1975
Mazda RX-7 Collection No. 1 1978-1981
Mercedes Benz Cars 1949-1954
Mercedes Benz Competition Cars 1950-1957
Mercedes Benz Cars 1954-1957
Mercedes Benz Cars 1957-1961
Mercedes 190 & 300 SL 1954-1963
Mercedes 230/250/280SL 1963-1971
Mercedes Benz SLs & SLCs Gold P. 1971-1989
Mercedes S & 600 1965-1972
Mercedes S Class 1972-1979
Mercury Muscle Cars 1966-1971
Metropolitan 1954-1962
MG Gold Portfolio 1929-1939
MG TC 1945-1949
MG TD 1949-1953
MG TF 1953-1955
MGA & Twin Cam Gold Portfolio 1955-1962
MG Midget Gold Portfolio1961-1979
MGB Roadsters 1962-1980
MGB MGC & V8 Gold Portfolio 1962-1980
MGB GT 1965-1980
Mini Cooper Gold Portfolio 1961-1971
Mini Muscle Cars 1961-1979
Mini Moke Gold Portfolio1964-1994
Mopar Muscle Cars 1964-1967
Morgan Three-Wheeler Gold Portfolio 1910-1952
Morgan Plus 4 & Four 4 Gold P. 1936-1967
Morgan Cars 1960-1970
Morgan Cars Gold Portfolio 1968-1989
Morris Minor Collection No. 1 1948-1980
Shelby Mustang Muscle Portfolio 1965-1970
High Performance Mustang IIs 1974-1978
High Performance Mustangs 1982-1988
Oldsmobile Automobiles 1955-1963
Oldsmobile Cutlass & 4-4-2 1964-1972
Oldsmobile Muscle Cars 1964-1971
Oldsmobile Toronado 1966-1978
Opel GT 1968-1973
Packard Gold Portfolio 1946-1958
Pantera Gold Portfolio 1970-1989
Panther Gold Portfolio 1972-1990
Plymouth Barracuda 1964-1974
Plymouth Muscle Cars 1966-1971
Pontiac Tempest & GTO 1961-1965
Pontiac Muscle Cars 1966-1972
Pontiac Firebird & Trans-Am 1973-1981
High Performance Firebirds 1982-1988
Pontiac Fiero 1984-1988
Porsche 356 1952-1965
Porsche 911 1965-1969
Porsche 911 1970-1972
Porsche 911 1973-1977
Porsche 911 Carrera 1973-1977
Porsche 911 Turbo 1975-1984
Porsche 911 SC 1978-1983
Porsche 914 Collection No. 1 1969-1983
Porsche 914 Gold Portfolio 1969-1976
Porsche 924 Gold Portfolio 1975-1988
Porsche 928 1977-1989
Porsche 944 1981-1985
Range Rover Gold Portfolio 1970-1992
Reliant Scimitar 1964-1986
Riley Gold Portfolio 1924-1939
Riley 1.5 & 2.5 Litre Gold Portfolio 1945-1955
Rolls Royce Silver Cloud & Bentley 'S' Series Gold Portfolio 1955-1965
Rolls Royce Silver Shadow Gold P. 1965-1980
Rover P4 1949-1959
Rover P4 1955-1964
Rover 3 & 3.5 Litre Gold Portfolio 1958-1973
Rover 2000 & 2200 1963-1977
Rover 3500 1968-1977
Rover 3500 & Vitesse 1976-1986
Saab Sonett Collection No.1 1966-1974
Saab Turbo 1976-1983
Studebaker Gold Portfolio 1947-1966
Studebaker Hawks & Larks 1956-1963
Avanti 1962-1990
Sunbeam Tiger & Alpine Gold P. 1959-1967
Toyota MR2 1984-1988
Toyota Land Cruiser 1956-1984
Triumph TR2 & TR3 1952-1960
Triumph TR4, TR5, TR250 1961-1968
Triumph TR6 Gold Portfolio 1969-1976
Triumph TR7 & TR8 Gold Portfolio 1975-1982
Triumph Herald 1959-1971
Triumph Vitesse 1962-1971
Triumph Spitfire Gold Portfolio 1962-1980
Triumph 2000, 2.5, 2500 1963-1977
Triumph GT6 1966-1974
Triumph Stag 1970-1980
TVR Gold Portfolio 1959-1990
VW Beetle Gold Portfolio1935-1967
VW Beetle Gold Portfolio1968-1991
VW Beetle Collection No.1 1970-1982
VW Karmann Ghia 1955-1982
VW Bus, Camper, Van 1954-1967
VW Bus, Camper, Van 1968-1979
VW Bus, Camper, Van 1979-1989
VW Scirocco 1974-1981
VW Golf GTI 1976-1986
Volvo PV444 & PV544 1945-1965
Volvo Amazon-120 Gold Portfolio 1956-1970
Volvo 1800 Gold Portfolio 1960-1973

BROOKLANDS ROAD & TRACK SERIES

Road & Track on Alfa Romeo 1949-1963
Road & Track on Alfa Romeo 1964-1970
Road & Track on Alfa Romeo 1971-1976
Road & Track on Alfa Romeo 1977-1989
Road & Track on Aston Martin 1962-1990
R & T on Auburn Cord and Duesenburg 1952-84
Road & Track on Audi & Auto Union 1952-1980
Road & Track on Audi & Auto Union 1980-1986
Road & Track on Austin Healey 1953-1970
Road & Track on BMW Cars 1966-1974
Road & Track on BMW Cars 1975-1978
Road & Track on BMW Cars 1979-1983
R & T on Cobra, Shelby & Ford GT40 1962-1992
Road & Track on Corvette 1953-1967
Road & Track on Corvette 1968-1982
Road & Track on Corvette 1982-1986
Road & Track on Corvette 1986-1990
Road & Track on Datsun Z 1970-1983
Road & Track on Ferrari 1975-1981
Road & Track on Ferrari 1981-1984
Road & Track on Ferrari 1984-1988
Road & Track on Fiat Sports Cars 1968-1987
Road & Track on Jaguar 1950-1960
Road & Track on Jaguar 1961-1968
Road & Track on Jaguar 1968-1974
Road & Track on Jaguar 1974-1982
Road & Track on Jaguar 1983-1989
Road & Track on Lamborghini 1964-1985
Road & Track on Lotus 1972-1981
Road & Track on Maserati 1952-1974
Road & Track on Maserati 1975-1983
R & T on Mazda RX7 & MX5 Miata 1986-1991
Road & Track on Mercedes 1952-1962
Road & Track on Mercedes 1963-1970
Road & Track on Mercedes 1971-1979
Road & Track on Mercedes 1980-1987
Road & Track on MG Sports Cars 1949-1961
Road & Track on MG Sports Cars 1962-1980
Road & Track on Mustang 1964-1977
R & T on Nissan 300-ZX & Turbo 1984-1989
Road & Track on Peugeot 1955-1986
Road & Track on Pontiac 1960-1983
Road & Track on Porsche 1951-1967
Road & Track on Porsche 1968-1971
Road & Track on Porsche 1972-1975
Road & Track on Porsche 1975-1978
Road & Track on Porsche 1979-1982
Road & Track on Porsche 1982-1985
Road & Track on Porsche 1985-1988
R & T on Rolls Royce & Bentley 1950-1965
R & T on Rolls Royce & Bentley 1966-1984
Road & Track on Saab 1972-1992
R & T on Toyota Sports & GT Cars 1966-1984
R & T on Triumph Sports Cars 1953-1967
R & T on Triumph Sports Cars 1967-1974
R & T on Triumph Sports Cars 1974-1982
Road & Track on Volkswagen 1951-1968
Road & Track on Volkswagen 1968-1978
Road & Track on Volkswagen 1978-1985
Road & Track on Volvo 1957-1974
Road & Track on Volvo 1975-1985
R&T - Henry Manney at Large & Abroad

BROOKLANDS CAR AND DRIVER SERIES

Car and Driver on BMW 1955-1977
Car and Driver on BMW 1977-1985
C and D on Cobra, Shelby & Ford GT40 1963-84
Car and Driver on Corvette 1956-1967
Car and Driver on Corvette 1968-1977
Car and Driver on Corvette 1978-1982
Car and Driver on Corvette 1983-1988
C and D on Datsun Z 1600 & 2000 1966-1984
Car and Driver on Ferrari 1955-1962
Car and Driver on Ferrari 1963-1975
Car and Driver on Ferrari 1976-1983
Car and Driver on Mopar 1956-1967
Car and Driver on Mopar 1968-1975
Car and Driver on Mustang 1964-1972
Car and Driver on Pontiac 1961-1975
Car and Driver on Porsche 1955-1962
Car and Driver on Porsche 1963-1970
Car and Driver on Porsche 1970-1976
Car and Driver on Porsche 1977-1981
Car and Driver on Porsche 1982-1986
Car and Driver on Saab 1956-1985
Car and Driver on Volvo 1955-1986

BROOKLANDS PRACTICAL CLASSICS SERIES

PC on Austin A40 Restoration
PC on Land Rover Restoration
PC on Metalworking in Restoration
PC on Midget/Sprite Restoration
PC on Mini Cooper Restoration
PC on MGB Restoration
PC on Morris Minor Restoration
PC on Sunbeam Rapier Restoration
PC on Triumph Herald/Vitesse
PC on Spitfire Restoration
PC on Beetle Restoration
PC on 1930s Car Restoration

BROOKLANDS HOT ROD 'MUSCLECAR & HI-PO ENGINES' SERIES

Chevy 265 & 283
Chevy 302 & 327
Chevy 348 & 409
Chevy 350 & 400
Chevy 396 & 427
Chevy 454 thru 512
Chrysler Hemi
Chrysler 273, 318, 340 & 360
Chrysler 361, 383, 400, 413, 426, 440
Ford 289, 302, Boss 302 & 351W
Ford 351C & Boss 351
Ford Big Block

BROOKLANDS RESTORATION SERIES

Auto Restoration Tips & Techniques
Basic Bodywork Tips & Techniques
Basic Painting Tips & Techniques
Camaro Restoration Tips & Techniques
Chevrolet High Performance Tips & Techniques
Chevy Engine Swapping Tips & Techniques
Chevy-GMC Pickup Repair
Chrysler Engine Swapping Tips & Techniques
Custom Painting Tips & Techniques
Engine Swapping Tips & Techniques
Ford Pickup Repair
How to Build a Street Rod
Land Rover Restoration Tips & Techniques
MG 'T' Series Restoration Guide
Mustang Restoration Tips & Techniques
Performance Tuning - Chevrolets of the '60's
Performance Tuning - Pontiacs of the '60's

BROOKLANDS MILITARY VEHICLES SERIES

Allied Military Vehicles No.1 1942-1945
Allied Military Vehicles No.2 1941-1946
Complete WW2 Military Jeep Manual
Dodge Military Vehicles No.1 1940-1945
Hail To The Jeep
Land Rovers in Military Service
Off Road Jeeps: Civ. & Mil. 1944-1971
US Military Vehicles 1941-1945
US Army Military Vehicles WW2-TM9-2800
VW Kubelwagen Military Portfolio1940-1990
WW2 Jeep Military Portfolio 1941-1945

21123

Contents

5	**Pre-Production Mokes**			
6	Mini and the Moke	Wheels & Tracks	No. 19	1987
14	Rumblings	Motor Sport	Feb.	1963
15	The Schizophrenic Moke	Autocar	Feb. 15	1963
16	Two-Pack Workhorse	Alternative Cars	Nov-Dec.	1982
19	**English Production Mokes**			
20	Payload Tests - Austin Mini-Moke	Payload	Jul-Aug.	1965
24	The Mini-Moke	Motoring Which	Oct.	1967
27	Mini-Moke	Car	Sept.	1965
28	Preposterous Mini-Moke	Popular Mechanics	Oct.	1965
30	The Golf Caddy of the 70's	Motoring	Sept.	1964
31	The Mini Moke Is No Joke	High Performance Cars	Oct.	1967
34	The Tailless Donkey	Motoring	Aug.	1965
36	McMoke	Austin	Dec.	1966
37	Mini Moke	Drive	New Year	1968
39	**Australian Production Mokes**			
40	Moke - runabout for the "IN" crowd?	Australian Motor Manual	March	1967
42	Motoring - Moke Style	Australian Motor Sports & Automobiles	March	1970
44	Running A Moke	Hot Car	March	1975
47	Making Tracks	Classic and Sportscar	July	1983
50	We Poke the Moke	Bushdriver	No. 3	Vol. 1
55	Does the Moke need 4WD for Overlanding?	Overlander	May	1979
59	Almost (And Maybe) The 4x4 Moke	4x4	Dec.	1982
63	**Portuguese Production Mokes**			
64	Moke Californian Tested	Alternative Cars	Jan.	1984
70	Go with Mokes!	Auto Express	June 23	1989
72	A Man and a Moke	Autocar	Nov. 26	1986
75	The Moke Factory	Mini Moke News	Oct.	1988
77	**Cagiva Production Mokes**			
77	Italians Revive Mini-based Moke	Autocar & Motor	April 25	1990
77	Mini Moke Production	Your Classic	June	1990
78	Mini Moke	Auto Express	April 9	1993
81	**Mokes Off Road**			
81	In The Rough	Worldwide	Aug-Sept.	1963
82	Unsuitable for Motors	Motoring	July	1965
84	Mighty Mokes Take on the Outback	Australian Adventures		1984
92	The World's Most Remote Off-Roader	Off-Road	April	1989
97	**Racing Mokes**			
98	Mudplug '68	Austin	Sept.	1968
102	No Moke Without Mire	Motor	Dec. 9	1967
104	Mini-Cooper S Moke	Autocar	April 18	1968
106	Here Come de Coke Moke	Bushdriver	No. 3	Vol. 1
110	RaceMoke	The Moke Chronicles	Vol. 4 No. 3	1987
111	**The Cult Moke**			
113	Demobbed	Classic and Sportscar	Dec.	1986
116	Mini Moke	BMC User	Feb-March	1968
118	Moking Around	Austin	Aug.	1968
121	Moke A La Mode	Car	July	1966
125	**Unusual Mokes**			
126	Mokes	Autocar	May 25	1967
132	Moke - Down Under	Transport Efficiency	June-July	1967
134	Moke at Large	Transport Efficiency	Feb-March	1967
137	**Adventures by Moke**			
138	Huntin' Fishin' Shootin'	Australian Motor	Jan.	1967
142	From Sea to Shining Sea	Road & Track	June	1991
146	Mini-Moke To Wild Wales	Cars & Car Conversions	Feb.	1967
147	A Moke at 8000 feet!	Modern Motor	Dec-Jan.	1968
155	**Advertising the Moke**			
169	**Moke Restorations**			
169	Making the Moke of it!	Practical Classics	July	1990
174	Donkey Work on a Mini Moke	Practical Classics	Oct.	1992
177	Mini Moke: After the New Gets Put On....	Road & Track	Aug.	1973

Foreword and Acknowledgements

The Mini Moke has to be one of the most unusual vehicles ever produced by the British motor industry. A military jeep that became a symbol of fashion, it has a fascinating history.

A jack-of-all-trades, the Moke's versatility has thrown it into many different spheres. Not quite a Land Rover, yet it will happily tackle fields, hills and green lanes. Not quite a sports car, but you can still drive along with the wind in your hair enjoying the sensation of speed. Not quite a truck, but Mokes have carried some pretty peculiar loads in their time. And if it rains, well that's all part of the fun!

The little donkey (moke *n* [origin unknown] 1 *slang Brit:* Donkey 2 *slang Austral:* Nag) has been kicked in the teeth many times, but has refused to die. It is to its credit that this endearing little vehicle has been in production for 30 years. It is loved by its owners and supported by Clubs around the world, dedicated to both the preservation and the use for all its designed purposes.

In this book you will find articles showing the many facets of the Moke, and chronicling its remarkable history from its origins at the drawing board of Sir Alec Issigonis at BMC's Longbridge works, through it's Australian and Portuguese years to its current promoters, Cagiva of Italy. On the way we shall stop to examine the Twini, the Californian, and the sports car that never quite was.

We are indebted to the publishers of 4x4, Alternative Cars, Austin, Australian Adventures, Australian Motor, Australian Motor Manual, Australian Motor Sports & Automobiles, Autocar, Auto Express, BMC User, BMC World, Bushdriver, Car - South Africa, Car & Car Conversions, Cars, Classic and Sportscar, Drive, Hot Car, Modern Motor, Motor, Motor Sport, Motoring, Motoring Which, Off-Road, Overlander, Payload, Popular Mechanics, Practical Classics, Road & Track, Transport Efficiency, Wheels & Tracks, Worldwide and Your Classic for allowing us to include their interesting and copyright articles.

We are very grateful to Anders Ditlev Clausager and the rest of the team at the British Motor Industry Heritage Trust Archives for their assistance in the research for this book and loan of magazines. Thanks are also due to Colin Atkinson, Raymond Baxter, Paul Beard, Ken Butterfield, Sherry Chandler, Max Hora, Peter Jones, John Kirby, Peter Metcalfe, Ian O'Neill, R.W. Phillips, Graham Robinson, John Tompkins, Andrew Turford, and Andrew Vanko for their help with articles, information, research and photographs. The Mini Moke on the front cover is owned by Ken James. The Mokes on the rear cover by Graham Robinson, Thomas Taylor, John Pearce and Barry Clough. The photographers were Neil Tuson and Ian Hodgson

The first edition of this book in 1989 coincided with the end of the line for the Austin Mini Moke. It was remarked then that the Mini Moke had always bounced back before, and that time was no exception as production in Portugal restarted in 1991 under the ownership of the Italian company, Cagiva. As this book goes to press, production is once more halted as the tooling and production track transfers to head office in Italy. Current plans are to restart production for 1995 with a 1275cc engine and catalytic converter. The previous resiliance of the Mini Moke leads us to hope that this plan will bear fruit and we will still be celebrating the Moke for many years to come.

Tim Nuttall
Editor, Mini Moke News
Mini Moke Club
December 1993

Pre-Production Mokes

The Mini Moke was conceived and designed in the late 50's at the same time as the Mini Saloon. Developed to meet a need in the British Army it went through a number of different prototype models over a period of five years, in BMC's attempts to persuade the army to take it.

The photographs in the article that follows, *Mini and the Moke*, show this development phase from the original basic buckboard (1959), through a shorter wheel base model with different styling (1962, and pictured below), to the immediate predecessors (1963) of the production model. The article (which is the most complete history of the Moke ever produced in one article) also chronicles the development of perhaps the most interesting prototype of all, the Twin Engined 4x4 Moke (1962). This prototype, of which only a handful were made, received world wide publicity after it was shown to the press in January 1963, during an exceptionally severe winter, sporting snow ploughs and hay bail carriers. A couple of the resulting articles are reproduced here together with a description of the Citroën equivalent, a twin engined 2CV known as the Sahara. This did have a small production run of around 600, but the twin engined Moke however, was again rejected by the army and the project was shelved.

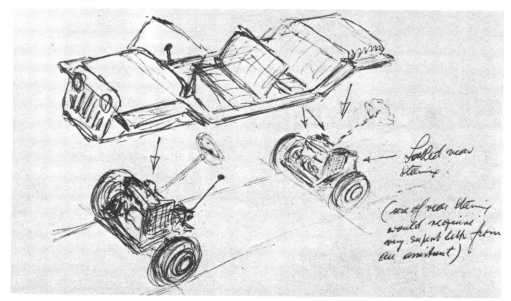

MASTER TOUCH.—This rough sketch by Mr. Alec Issigonis shows the twin-engined Mini-Moke conception. His pencilled note on the rear assembly reads "use of rear steering would require very superb help from an assistant."

THE MOTOR
January 23 1963

THE MINIMOKE

THE MOTOR February 21 1962

CONSIDERABLE interest was aroused last week by reports in the Press of a "Minimoke," the military version of those highly successful B.M.C. twins, the Austin Mini and the Morris Mini-Minor. This vehicle is not in fact new, writes Harold Hastings. It has been under development almost from the inception of the Mini design and was shown at a military review some two years ago.

The general layout follows that of the Mini, with the engine placed transversely at the front and the running gear grouped on two chassis sub-frames—the front carrying the engine, the rubber-cone independent front suspension and the steering units, with the second sub-frame bearing the independent rubber-cone rear suspension. This layout makes it easy to use the basic components with any suitable main structure and the Moke has a general-purpose body designed expressly for possible military use.

Another difference is that the engine is of greater capacity (948 c.c. instead of 848 c.c.) to enable the necessary power to be obtained with low octane fuel.

It will be at the Amsterdam Commercial Motor Show which opens tomorrow.

MINI MOKE *AUTOCAR, 21 August 1964*

PRODUCED ORIGINALLY as a military vehicle, the front-wheel-drive Mini Moke, either as a Morris or an Austin, is now in production for general sale. It has the standard Mini 848 c.c. power unit, a fully rustproofed, unit construction steel body with a very low centre of gravity, and is fitted with a vinyl-treated fabric tilt with detachable hood tubes. The windscreen can be folded forward, or removed completely. The rack and pinion steering has $2\frac{1}{3}$ turns from lock to lock and most chassis details are similar to those of a normal Mini saloon, including suspension and brakes.

A variety of uses suggest themselves, including as a beach car, hotel taxi, factory transport, site survey vehicle and even a golf buggy. Fitted with a driving seat only, the basic price is £335, total with purchase tax £405 7s 1d. Extras available include a front passenger seat, two rear seats, laminated windscreen, sump guard and Dunlop Weathermaster tyres.

The experimental Mini-moke provides seating for four, excellent visibility and as much fresh air as anyone could want.

The original Mokes were very basic runabouts, with bench seats. Windscreen was a later addition. On the right are the first-ever pictures to reach the public eye, in 1960.

Mini and the Moke

A survey of BMC/Leyland's military and civilian Buckboards

In automotive history the world has witnessed several more or less successful attempts to conceive and build utilitarian cars for military use on the basis of contemporary mass-production models. Early examples were the Ford T with bucket seats and the old Austin Seven scout cars of the British Cavalry. Later there were the VW Kübel and its post-war version, the Type 181, as well as DAF's 66YA and the East German Grenz-Trabant. In France there were the Citroën Mehari and a cut-down variant of Renault's R4. Britain's post-war contribution was the Moke (= donkey).

Most of these vehicles were also

Designer Alec Issigonis and early Minis.

marketed commercially, for utility and recreational purposes; 'fun cars' perhaps being the best description. Offered in sundry pretty colour schemes and with a variety of optional extras, the idea usually was to cash in on their availability and/or to offset production costs. A few, like the VWs, had originally been designed with military use as the prime objective.

The Moke was conceived in the late 1950s, concurrently with BMC's famed Mini, under the leadership of Mr (later Sir) Alec Issigonis. In the Austin Design Office, the Mini project was known as ADO15 and the Moke variant was developed to meet a British military requirement. It is uncertain which design came first but they were both tremendously successful, both still

On the first Mokes, battery, spare wheel, fuel tank and pump were in the rear. Damage to Royal Marines 07RN49's seat backs was caused by a Whirlwind helicopter — which had just delivered it — accidentally sitting on it!

14BT18 when new, with full and cut-down steering wheel, seen at Austin's Longbridge, Birmingham, works in 1959.

RAF Moke (27AE03) dwarfed by Army Antar and, right, a 1962 military prototype during a demo at Umtali in August 1962, being lifted by a chopper of the Royal Rhodesian Air Force.

being in production to this day. Of course, many detail alterations and improvements have been made but in general appearance little has changed.

The Minis were originally known as the Austin Seven (or Se7en) and the Morris Mini-Minor but the 'mini' name became so universal that after 1962 BMC called them all Minis and in 1969 Mini became the marque name in its own right. The 5,000,000th was built in February 1986. In addition to the saloon there were estate cars, vans and pickups (and even Riley and Wolseley derivatives).

When Issigonis masterminded the Mini, he already had some well-known concepts to his credit, notably the Morris Minor and the four-wheel drive Nuffield Gutty which led to the production of what became the Austin Champ. His greatest achievement was the positioning of a 'normal' four-cylinder engine cross-wise — 'east-west' — driving the front wheels via a unitary gearbox/final drive system and all sharing the same lubricant, engine oil. We say 'normal' engine because before the war the German Auto-Union's DKW division had mass-produced cars and vans of similar configuration but these had comparatively narrow two-cylinder two-stroke power units.

Issigonis did it again with the BMC 1100 and 1800 as well as the Maxi and others and in later years the design was copied by manufacturers all over the world. Today, a transversal engine with front-wheel drive is as commonplace as the traditional layout.

The British Army in the late 1950s wanted a small, lightweight, air-portable (36 in a Beverley transport), economical, cheap runabout. Their lightest tactical vehicle at the time was the Austin Champ, which had all but the above characteristics. Of course, the lightweight could never replace the ¼-ton 4×4 as such, but the Champ's

The 1962 model had a shorter wheelbase and a very different body shell, with hinged windscreen. Engine was 950 cc. BMC pamphlet 2105/A said it was 'available to Government Departments for quantity production only'. Certain production variations were offered to suit export requirements, but there were no takers. This left-hand drive example appeared at the FVRDE/SMMT exhibition in 1962.

Concurrently with the short wheel base model, BMC refined the original 80-inch (left). Further developed, the 1963 version (right) became the immediate predecessor of the 1964–68 production model. It was tested by the AA but for them, not surprisingly, the Mini-Van was found to be a better proposition.

The first Twinni-Moke, constructed in 1962, was based on the early style Buckboard design and had the complete power/drive train duplicated at the rear.

The original Twinni-Moke being demonstrated to the press in January 1963. Fitted with a tiny snow plough — hardly visible here — and carrying a load of hay, it is driven by Mr W. J. (Jack) Daniels, BMC's 'A' Group Projects Superintendent, who, together with John Sheppard, produced the prototypes. The separate gear lever for the hind power unit was on the right, which, contrary to expectations, worked quite well.

capabilities were hardly ever utilized to the full, certainly not in peacetime conditions. In that respect the Moke, as an additional model, made a lot of sense.

The Mini's power train (with the engine bored out to 950 cc) and Moulton rubber suspension were left largely unaltered and bolted to a punt-style open body structure which was new and totally different from the car (and van) with the exception of a couple of steel pressings in the bulkhead area. BMC called this utility body design Buckboard.

Half a dozen prototypes were hand-built in mid-1959 — prior to the saloon being launched — and in 1960 supplied to the Fighting Vehicles Research and Development Establishment (FVRDE) at Chertsey in Surrey for rigorous tests. One or two went to North Africa for a three-month evaluation period. The main shortcoming turned out to be the chronic lack of ground clearance, which always remained the vehicle's greatest defect. (On the rear-drive VW Kübel this had been easy to rectify by installing gear casings at the ends of the swing axles; it also had much larger wheels.)

Of those original Mokes, at least four can be readily identified, mainly from pictorial evidence. At least three were tested by the Army: 14BT17, 14BT18 and 20BT30. All of these have survived, one in the Museum of Army Transport at Beverley, the others in private hands. The Royal Marines tested one for helicopter lifts and drops; it carried registration 07RN49. There was also one with the RAF: 27AE03.

The Army tests were terminated in 1963, when 14BT17 and 20BT30 were slated for the School of Transport Vehicle Museum. 20BT30 was used on several occasions in events such as the BBC-TV Autopoint competitions and also appeared at Army displays.

The Royal Navy decided to standar-

Registered 2H0669 and 2H0670, the US Army in 1964 tested a standard and a twin-engined model, the latter with 1100 units and 12-inch wheels. The two gear sticks were now one behind the other and joined by a coupling (which could be disconnected when using only one engine). The tail end was also different and several were made. One has survived in the BL Heritage Collection.

The standard British civilian Mini-Moke of 1964–68 attracted much attention from the accessory industry. Here we see twin front tyres (5.20-10), electric winching kit, hardtop and, below, the PT-dodging 'Mini-Truk' offered by Bradburn & Wedge of Wolverhampton in 1965. In its standard form, the Mini-Moke became popular with the 'Chelsea set'. The Austin and Morris editions were designated A/AB1 and M/AB1 respectively. Individual customizations included 1,275-cc Mini-Cooper S engine installations and even six-wheelers.

dize Citroën 2CV pickups instead of the Moke, especially for use on and off aircraft carriers, but later they also used some standard Mokes.

The Army's complaints about lack of ground clearance — particularly below the engine — were rectified by BMC to the best of their ability. The result was a completely new Moke which first appeared in early 1962. The wheelbase was reduced to 72.5 inches (7.5 inches less than the saloon and the early Mokes) and the ground clearance was improved somewhat — to the maximum possible — by inserting 'packing' between suspension units and bodyshell. A hefty metal guard below the engine protected the sump, which also contained the gear train.

The Moke's lack of four-wheel drive was not its major shortcoming. In fact, with most of its weight on the driving front wheels, the off-road performance was not bad at all. A diff lock would have been an asset, but it was hardly fair to expect the off-road performance of, say, a Land-Rover from a tiny vehicle — derived from a small 4×2 car — that was never required to do the same work as its larger brothers. Cus-

tomers who did were clearly on the wrong track. They could and should (and often did) buy a Haflinger!

However, in 1962 BMC did start experiments with a 4×4 Moke by the comparatively cheap and easy expedient of 'repeating' the power train at the rear (à la Tempo G1200 of the '30s). This eight-cylinder 1696-cc 'Twinni-Moke' had tremendous acceleration and hill-climbing abilities, the ascent of a one-in-two gradient being quite feasible. It received world-wide publicity when demonstrated to the Press during January 1963, in the midst of an exceptionally severe winter, carrying bays of hay to isolated cattle — and pushing a snow plough. We well recall driving this

vehicle a little later in a treacherous gravel pit and being very impressed (with a gear lever for each hand a feast for fervent gear shifters!).

All being said and done, the Twinni-Moke did not make a lot of sense, however, except perhaps as a fun machine. It was weird and complex for what it was and could carry only two passengers. Yet, the US Army acquired one for tests, a left-hand drive model with two 1100-cc engines, giving a combined power output of 110 bhp. It went to the Tank-Automotive Command in Warren, Michigan, along with a standard 850-cc model. (Where did they end up?)

In 1966 BMC introduced an experimental 4×4 with (single) 1100 engine, known as the Ant, but this was a rather divergent vehicle and therefore another story.

Meanwhile, the British Army had also rejected the 'short' 1962 950-cc model (which looked very different from previous as well as later models) and in 1963 BMC revived and developed the 80-inch wheelbase type, with many refinements. This became the pilot model for the civilian Mini-Moke,

Sir Alec Issigonis with some of his creations and an A/AB1 converted for ambulance use in New Guinea, both in the mid-1960s.

After Moke production had moved to down under, the Australian Army tested a handful for use in tropical conditions. The little trailer proved unreliable and was rejected. Note the special stack type air filter (which was not standardized either). When bogged, the Moke could easily be extracted by muscle power.

which went into series production, with 850-cc engine, in 1964. It was available either as Austin or Morris (depending on the dealer who sold it). The Royal Navy bought at least one, a Morris, in 1965. Registered 10RN30 it was demobbed ten years later and sold at Ruddington.

The civvy Mini-Moke was in production in Britain for only four years. A total of 14,518 were made, of which only ten per cent (1,467) were sold on the home market. The main reason for its early demise was the Treasury's ruling that the Moke was a car, attracting purchase tax. In 1967, its last full year in Britain, the Moke with driver's seat only cost £335, to which the taxman added £78 PT. For the other three seats (if required), with grab handles, the additional cost was just over £25. Peanuts by today's standards, of course, but it was very different then.

In October 1968 British Leyland chief Sir Donald Stokes announced that Moke production would be transferred, in its entirety, to Australia. This was to add 14 years to the vehicle's production life, but made it — as an import — much more expensive in its country of birth, where several firms introduced DIY kits for look-alikes (the mechanicals presenting no problems!).

Back in the early 1960s, BMC had demonstrated the Moke to the armed forces of several Commonwealth nations, including Australia, where in 1963 a 1,000-cc version had been tried by Army Headquarters. The Aussies had the same objections as their British counterparts earlier. But on March 31, 1966, the Mini-Moke — assembled in the company's Australian Victoria Park, Sydney factory — was launched as a civilian vehicle, with a price tag of A$1,295, including tax. This was much the same vehicle as in Britain but changes had been made to adapt it better to local conditions, including 1,100-cc vs 850-cc engine, lower gearing, stronger full-width sump guard, better front seats, dual wipers and a parcel shelf on each side of the instrument cluster. The top bow and the side curtains were also improved and all electrics and plumbing were fitted higher, out of harm's way.

Soon after, all production was moved to Enfield, New South Wales, and the Army looked at the Moke again: they procured eight units (registered 176891, 92, 94–99), with 13-inch wheels and 1,100-cc engine, together with four matching BMC-built single-axle trailers, for tropical trials. Two vehicles and one trailer were assigned to the Tropical Trials Establishment (TTE) for a twelve-month period, starting in September 1969. The remainder went to Papua New Guinea Command (PNG Comd). Of many defects found, most were rectified and modifications introduced in production on all vehicles. Greatest problems were experienced with the exhaust system and the shock absorbers. Other components that needed attention included the spare wheel securing mount, the windscreen (to be fitted with laminated glass) and the wiring (fuse to be incorporated in tail and side light circuit). The provision of various items (e.g. a reversing light, tow rope, plugs for body drain holes, etc.) were suggested and the special stack-type air cleaner was deemed superfluous. Access, particularly with the top raised, was found extremely poor and the seats being non-adjustable and non-collapsible were not liked either.

Altogether, however, the Australian Army was pleased enough with their new 'Truck, Lightweight, ¼-ton, CL (BMC Moke)' to order an initial 111 units, soon followed by some 400 on a period contract. The ¼-ton trailers, incidentally, proved too small, flimsy and troublesome and were rejected.

Meanwhile, Moke production in the Antipodes went well and many were

Australian civilian production model of 1970 (left) differed in many details from the British ones, partly as a result of the Army's criticisms, and (right) one of the first series-produced military models, as seen at the Zetland factory by Laurie Wright in early 1973. Wheelbase was 83 inches, tyre size 5.60-13. Many, if not most, of the Army Mokes have now been demobbed.

Left: Colombia was one of the many export markets of Leyland Australia. Right: Mokes could be stacked in pairs for shipment, thus lowering freight costs (by A$65 per vehicle to Caribbean ports, for example).

exported, reportedly to over 30 countries. Military customers, other than the Army, included the Australian Air Force and Navy and the New Zealand Navy.

Civilian Mokes, like the military ones, now had the 13-inch wheels and tyres, increasing the ground clearance to a more acceptable 8.25 inches. Track was also wider than on the British model and there were numerous other detail improvements. The 1,275-cc Mini engine was optional (and standard for the USA, fitted with exhaust emission control).

The Australian Army Moke differed from the civilian YDO 18 Mk II in several respects, for example: unit/formation sign holders, front and rear; thermal transmitter in engine head, operating a warning light on the facia 20 °F before boil; two-speed wipers; electrically-operated windscreen washers; fire extinguisher under parcel tray; 50-mm tow ball attached to rear bumper for trailer, with 12-pin socket for trailer wiring; 'throw-in' full-width rear cushion, removable for carrying stores; better (Vinyl/cotton) seating material; line fuse added to tail light circuit. There were about half a dozen more.

The following features were reported as being among the Moke's attractions, off-setting some of the drawbacks: lightness (two men could lift the rear end, three the front); traction, approach and departure angles, ride and roadholding; low overall height (41.5 inches to top of steering wheel).

By the early 1970s it looked as if the Australian Army's ageing Land-Rover fleet would be replaced by two new categories: the Moke for the light work and a newly-developed purpose-

Moke with Australian Army IHC Mk V 6×6 dump truck, November 1974. (Laurie Wright.)

Aussie truck version of the RAAF, used for guard dog transport, seen at Richmond Air Display, March 1979. Note spotlight and queer type tyre treads. (Laurie Wright.)

Another Australian derivative was the top-notch Californian, with special wheels, bull bars and other embellishments as standard. This specimen (in Tasmania, 1985) sports a polyester hardtop.

Royal New Zealand Navy Mokes, normally carried aboard HMNZS Canterbury, left (Norm Weeding) and Waikato, right (Laurie Wright), both snapped on Australian soil in 1978/79. Colour scheme was blue. Note sling flanges on hubs.

designed 1-ton GS truck (with forward control and four-wheel drive) for the bigger jobs. It all worked out differently. The 4×4s, which were prototyped by Ford and International, were tested but not adopted. Land-Rover reigned supreme, although, in practice, the new Series 3 was not liked at all.

In January 1975, Leyland Australia introduced a little truck version of the Moke with flat-floor Tray top body and a few years later the Californian, which was the basic Moke smartened up with white spoked wheels, smaller steering wheel, front and rear bull bars and other refinements. They even designed a 4×4 version (with one engine this time).

1982 saw the end of the Aussie Moke. The production lines were to be used for assembling Peugeot cars. . . .

But the little donkey wasn't dead. Manufacture was transferred once again, back to Europe! Leyland's factory in Setubal, Portugal, became the new producer. Now renamed Austin Rover Portugal, they soon got the hang of it and produced nearly 400 units in 1985, mostly — 304 — for export. About 20 of these found their way to the UK, via a private importer (Dutton) in Sussex. For 1986 the aim was to produce 1,000 Mokes of a revamped type, known as Moke 86, which used many components — including 12-inch wheels and disc/drum brakes — of the current-production Mini saloon. Some 90 per cent was destined for export, mainly onto the French market, the islands of the Caribbean and the Indian Ocean taking the balance. Moke 86 differed in many details from its predecessor, except in overall appearance.

So, for more than a quarter of a century now, both the Mini and the Moke have been with us and it looks as if both of them will be continued in production for some time yet; they are that appealing! □

Re-engineered 1986 Moke, made by Austin Rover's subsidiary in Portugal.

Back to the beginning: two of at least four survivors from the original batch of Buckboards: Tony Oliver's (serial SPL468/AM), the only one still with the original engine (SPL462-A-950 cc) (Photo: David Shephard) and, right, 14BT17 (serial SPL446/AM) which is in the Museum of Army Transport at Beverley. AM stands for Austin/Morris, SPL for Special (prototype). 20BT30 (SPL453/AM) and 14BT18 (SPL466/AM) are owned by Ian Marshall of Portsmouth and Graham Robinson of Bournemouth respectively.

RUMBLINGS

A TWIN-ENGINED B.M.C. MOKE 4×4

One Friday in January the Editor was the recipient of a mysterious telephone call from Longbridge, Reg Bishop of B.M.C. Publicity requesting his company at the Austin works by 12 noon on the Monday. We agreed to go, imagining that a B.M.C.-Rolls or a new Cheshunt-challenging rally car was to be released.

It seemed rather tactless to visit the makers of front-engined f.w.d. mini cars in a rear-engined vehicle, but the Morris 1100 being sick of a stiff gear-change we had to go in a Renault R8, which took us comfortably, warmly and in safety over the slippery snow-covered roads to Birmingham.

We left at 8.30 a.m., stopped for petrol, got lost as we do inevitably when trying to find the Austin factory from Redditch, but arrived soon after the appointed time, to find fellow journalists waiting about in the famous reception-cum-showroom for the main party to arrive. The latter had left Paddington around 9 a.m. and although snow had not fallen for some days, their train was running late. A photographer who had left London in his Ford at 10 a.m. arrived, but still we awaited the train party—which doesn't say much for Dr. Beeching's Railway Reformation.

Eventually, lunch was taken and we were told that we had been summoned hastily, while snow still remained, to see and sample B.M.C.'s latest baby, weaned by Alec Issigonis, their twin-engined 4×4 Moke.

This is a simple open-bodied vehicle with a normal 848-c.c. Mini unit at the front and a similar-engined sub-frame at the back but with the steering locked. Clutches and throttles are interconnected but each engine has its own ignition switch and gear-lever; thus there are two gear-levers, the normal central one and a r.h. one for the rear engine. This means that different ratios can be selected front and back, but on slippery ground no ill-effects result; what happens on dry roads or if the rear box alone is put into reverse we didn't discover.

Citroën, who have led in so many things, did this with two 2 c.v. power units long ago but they had less power to play with; the 1,696-c.c. twin-engined B.M.C. Moke totals 72 b.h.p. which should be impressive on the road, and, B.M.C. tell us, enables a 1 in 2 gradient to be vanquished.

The assembled journalists were provided with gum-boots and drove the thing on the Longbridge lawn, which was many inches deep in snow, but will never be the same again after the thaw! With straw bales on the back, a snow plough in front, the Dunlop Weathermaster-shod Moke got through deep snow in a continual under-steer on bends, and when it bogged down it could be reversed out. We were sorry, however, not to see it perform up slippery hills. (But must confess that when the engine of our new Morris 1100 packed up the other day we would have been very glad of another one in the boot . . . . And how pleasant to read in the hand book that when one engine breaks or wears out another is there for instant employment!) Some development to make gear selection easier seems desirable, but this vehicle had been completed in a week or so, to try it on virgin snows. It should be useful for military purposes, in undeveloped countries and on farms, etc., and Mr. Harriman told us that it will sell for about £475.

The publicity boys are already thinking of 1,100 c.c. power units giving a total of 110 b.h.p. but Issigonis says this is entirely unnecessary, the Moke has such admirable traction as it is. There was also speculation about a four-wheel-drive rally or racing car (why not for auto-cross?) on these lines but clearly the B.M.C. Moke's purpose in life is as a cross-country vehicle.

* * *

CITROEN'S VERSION

A week after we had sampled the B.M.C. Moke on the Longbridge lawn we drove to Slough, appropriately, this time in our inter-connected special suspension, front-drive Morris 1100 (now in good health again after the maladies recounted in "My Year's Motoring"—its stiff gear-change was explained as a remote control made to too-close tolerances) to drive the twin-engined 850-c.c. Citroën 2 c.v.

This was developed for Sahara pipe-line reconnaissance *and announced four years ago*. Outwardly, only the bonnet-mounted spare wheel, side fuel fillers and fan aperture in the boot-lid suggest the presence of a duplicate power unit in the tail.

The gearboxes, throttles and clutches are all inter-coupled, but

MOKE IN THE VIRGIN SNOW.—This superior B.M.C. baby has two engines.

there are separate ignition switches. Normally this 4-wheel-drive Citroën runs normally on its front engine, when its extra weight (14½ cwt., tanks full) renders it somewhat sluggish. But both engines can be used on the road, releasing a total of 28 b.h.p., when considerable additional performance is achieved. In this 4×4 form we experimented at Burnham Beeches and found adhesion up extremely steep snow-packed hills, quite phenomenal. Without anyone or any load in the back to aid wheel-grip the Citroën climbed gradients from rest that a normal car wouldn't look at after a fast run-in, and which were too slippery to walk up. It re-started effortlessly in 7 in.-deep frozen snow.

There is a new central floor gear-lever working in delightfully vintage fashion with very small movements, and a smaller lever which engages 4-wheel-drive. For single-engined motoring the front engine is used but if this fails the rear one can be brought into action after the front clutch has been permanently disengaged by means of a special hook-rod supplied in the tool-kit. The tyres are oversize Michelin "X," to give a good area of contact.

With its conventional closed 4-seater body, heater, and proper coupling of front and rear 4-speed all-synchromesh gearboxes, this Citroën is an advance on the B.M.C. Moke. It costs £912 17s. 1d. here, after import duty, sales and purchase-tax have been paid, but in France it costs the equivalent of £740. A top speed of 62 m.p.h. is claimed, with normal 2 c.v. economy when mono-motored, some 31½ m.p.g. using both 425 c.c. engines. Each has its own ignition light, fuel pump and 3¼-gallon fuel tank. Air-cooling, supple suspension, comfortable seats and high ground clearance are amongst its assets, and the low total power and properly synchronised road-wheel speed eliminates wheel-spin.

The abnormal English snows have certainly focused attention on cross-country cars. We understand that Innes Ireland has developed another, using a VW engine.

* * *

WINTER ACCESSORIES

Cleaning up the office after Christmas we came upon a variety of "goodies" sent in for comment. These can be listed as:

(1) A Charleson Components' "Fros-Free" jute-hessian cover for keeping frost off the windscreen, giving wheels a grip on slippery surfaces, protecting clothing when dealing with motoring predicaments, etc. Price, 12s. 6d.

(2) An Eolopress CO_2-charged tyre-inflator which also acts as a fire-extinguisher. Model 105 costs £6 10s. and will, they say, seal the beads of tubeless tyres. It works at 1,000 lb/sq. in. to start with but you have to weigh the cylinder accurately to discover whether it wants re-charging, which can only be done at approved agents. When full the bigger models will, they claim, inflate 16 5.00/5.20 × 14 tyres from flat.

(3) A big jar of United Lubricants' "Forlife" new anti-freeze claimed to be anti-corrosive but which changes colour if there is corrosion in the system. We offered this to a friend with an aluminium engine but he preferred to drain off. Ourselves, we use Castrol.

(4) A 100% Terylene tie incorporating a tiny head-on racing car in silver woven thread. Well, we met a gentleman in Germany who had looked everywhere for one—you get them from Les Leston Ltd, for 14s. 6d.

So far as the winter motoring aids go, we got by with a nylon tow-rope, a spade and our Dunlop tyre-pump.

The schizophrenic Moke

—and First Eight-Cylinder Austin

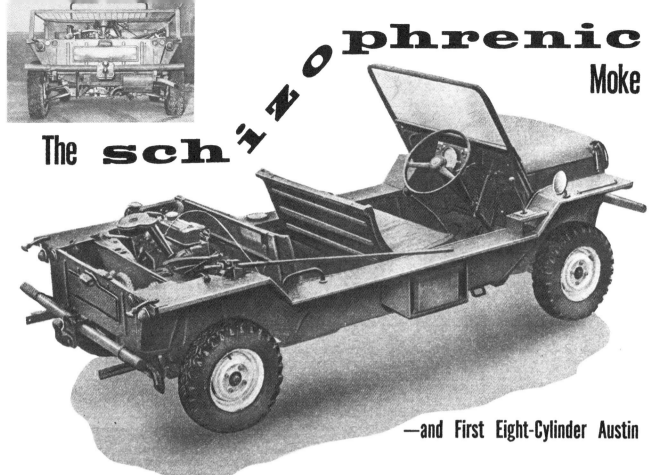

HITHERTO the many reports about the B.M.C.'s twin-engined version of the Mini-based military Moke have dealt only with its extraordinary abilities in deep snow. I have now been able to try it at speed on dry main roads, and to gather some impressions of the other side of its character. The only Moke which can kick with vigour at either end, it was described briefly in this journal on 11 January (page 72). The standard front-wheel drive version has a 948 c.c. low-compression engine, while the twin-engined prototype has this plus an ordinary 848 c.c. Mini unit at the back; the layout does not necessarily demand equal-sized engines, since there is no tractive interconnection. It probably *is* the first eight-cylindered Austin, but perhaps one should refer to it as a double-four.

For its first trial outings after being very hurriedly concocted in the experimental workshops at Longbridge, there had to be a riding mechanic sitting in the back to work the rear gear lever. Now there is a rod-and-bellcrank linkage bringing a short lever within reach of the driver's right hand. The front unit's lever is in the usual place in the middle of the floor. One accelerator and one clutch pedal serve both engines.

It was bitterly cold, but they gave me a dark blue duffle coat, no gloves and a pat on the back; and poor Jack Daniels, whose reward for helping to make the beast has been to sit and freeze in it beside numerous other journalists, came and did the same for me.

To set off from rest there are numerous possible combinations, such as: Front engine live and rear engine dead (or in neutral), or vice-versa; both engines live, both in first or second, or one in first and the other in second . . . and the next move offers a further wide choice. To begin with the temptation to play games with the two units in different gears is irresistible. Sometimes one engine can be heard singing up to high revs, but which one? There is a momentary hesitation while the frozen ears decide which frozen hand should make the next move.

Calling All Hands

From a purely practical viewpoint all this may sound rather nonsensical, if a lot of fun. Obviously there would need to be a further development bringing both gearboxes under the control of a single lever, whether by rods, cables or hydraulics. Yet the disadvantages of the independent levers are not so great as they might seem, and there is a lot to be said for retaining at least the means of dividing the controls. The main snag might seem that double the normal number of hand movements are called for. Yet with over 70 b.h.p. available for a very light weight, distributed evenly with the drive between front and rear, one can normally dispense with the first gears, start in seconds (two movements), then change up to third with one engine, top with the other and then top with the first—five movements instead of the conventional four.

Once on the move on a normal road, one has, in effect, a seven-speed box: 1 and 1, 1 and 2 (or 2 and 1), 2 and 2, 2 and 3 (or 3 and 2), 3 and 3, 3 and 4 (or 4 and 3), 4 and 4. For this count we disregard such two-step combinations as 1 plus 3 or 2 plus 4. To appreciate this more clearly, assume that you are attacking a hill with both gearboxes in top, and the speed is dropping; often it can be sustained by changing down to third with only one box. The main nuisance is when approaching a very slow corner in twin tops, and a double down-change has to be made. Strictly non-U are the simultaneous changes with both hands, which can be accomplished safely only on the straight—and not when anyone with a dark blue pointed hat is looking.

For cross-country work, especially when carrying a heavy burden, it might sometimes help to be able to apply extra torque to one end; for instance, when climbing a steep hill with most of the weight transferred to the rear wheels, one might do best with the front engine pulling in second, the rear in first.

Although the chassis structure of this first example, with its rather long wheelbase, lacks torsional rigidity, it handles quite well at speed and is very stable directionally. It also runs very smoothly indeed on all eight cylinders without any vibrating couples or other such nonsense. When one has experienced the Double Moke's acceleration, it is breathtaking just to imagine what a Mini saloon could do in competition with a really hot engine at each end, a power-weight ratio of over 200 b.h.p. per ton and four-wheel drive . . . which is already in prospect from more than one quarter. R. B.

Moke as a collectors' item today—at a gathering of military style vehicles.

Were the BMC Twini-Mokes the ultimate in unstoppable four-wheel-drives or were they just another also-ran 4WD line from Austin? Dave Shephard looks into this interesting variation of the Mini phenomenon.

Twenty-three years after its birth, the good ol' BL Mini is still a favourite. A classless car. And one which still sets the pace for today's box cars in performance, road-holding, economy and spaciousness.

As soon as the Mini car appeared in 1959 under the then BMC banner, it was apparent that the compact, self-contained, transverse front engine/transmission package would be popular for specialist uses. Build a vehicle, and the self-contained power package was already to bolt-in, together with the Mini's suspension all-round if needed.

DONKEY CAR
A van and pickup truck appeared a little later, in 1960 and '61 respectively. However, of special interest to off-road men was the utility Moke which appeared as a production model for public consumption in 1964. 'Moke' means donkey—very apposite for a new workhorse. But the public purchasers had other ideas for its use.

Four years of various Moke prototypes had preceded the final product's public appearance. These were produced mainly with an eye to the military markets. The Forces were looking for a light, general purpose transport suitable for air-dropping or slinging under a helicopter, and capable of being stacked for storage. A good idea. But the Moke never found great favour with the British Forces, though serious trials were carried out with various prototypes such as long and short wheelbase versions. Ground clearance was the major failing (6.4 inches on the production model) and probably the lack of four-wheel-drive also made the donkey 'lame'. But the lack of 4WD need not have been a problem, as we shall see.

GREAT MAN
The father of the Mini, the other famous British cars like the Morris Minor, Sir Alec Issigonis, is of Greek descent on his father's side, and came to England in 1923, driven by war from his homeland. Eventually he rose to become engineering chief at BMC, and was knighted in 1969.

With the petrol rationing of the Suez crisis in 1956 everyone was looking for a car that ran on nothing! Thus the Mini idea was born. It became known as ADO 15 (Austin Design Office project 15). Alec Issigonis, inspired by the FWD Citroen Traction Avante cars made from 1934, chose the FWD needed to five good passenger space. For the same reason a transverse engine was a must. To add to the compactness by installing the transmission in the crankcase, sharing the sump oil, was a brave move, but one which has been vindicated by the test of time.

DOUBLING-UP
Up to 1962 all the work on Minis and pre-production Mokes had been 4×2. Never a man to rest on his laurels, Alec Issigonis felt that already good

TWO-PACK WORKHORSE

Half the seats and twice the power. Doesn't look bad, either!

Two ignition switches and two gear levers show that it's no ordinary Mini-Moke.

performance of the Moke could be bettered with two engines—one to drive the front as normal and another pushing at the back. Within a space of a few weeks, a Moke had been re-worked and a second engine unit sat in the back with the wheels fixed in the straight-ahead position. The overall length remained the same, but the wheelbase was raised by 3 inches. Performance was said to be outstanding; the engine-over-drive-wheels—always good for traction—gave the Twini-Moke, as it was christened, plenty of thrust through the snow drifts in demonstrations during the bitter winter of 1963. In these trials a platform was added at the rear, over and behind the engine, to give some space for a payload, like fodder for animals isolated by the snow.

The two engines had a common clutch pedal, and the throttles were also linked, but the gear levers and ignition switches were separate. On this first model the front engine gear change was in the normal central position, and the rear change was rod-operated by a lever fixed to the offside body box. Quite a handful when doing a shift! It would have been possible, with the 'front' now at the back, to adapt the Moke to four-wheel steering, but nothing is known of such a development.

Of course, the Twini-Moke could be driven mono on the road, in the name of economy, simply by cutting the front or rear engine and selecting neutral. On later models the front and rear gear levers were cleverly engineered to sit together between driver and passenger, and were joined by a coupling. Removal of this coupling allowed separate operation of the gearboxes, and it was found that a difference front/back of one cog was acceptable. A skilled driver could thus select the degree of torque required at each end of the Twini.

The Twini-Moke was not, of course, the only light cross-country contender from the house of Austin: Champ, Gipsy and Ant are all 4WDs which tried to prosper in Land-Rover's shadow. But was the Twini-Moke the ultimate four-wheel-drive that even carried its own spare engine? Despite capable performance, the vehicle lacked useful cargo and passenger space, and its use as a practical solution for off-road life is questioned.

TRIALS IN THE U.S.
But interest there was aplenty. In 1964 the United States army tested Twini-Moke fitted with two 1098cc engines and 12 inch wheels (other Mokes had 10 inch wheels normally) for possible use as a ¼ ton passenger/cargo war car. The two engines naturally showed their worth, though surprisingly the top speed was only increased by 11mph over the laden mono-Moke. Laden, the US Twini could reach 67mph. Although it drove all four, the Twini was unlike a conventional 4WD vehicle, because only half the total engine power could appear at each end. Thus, if one end lost traction, the other engine was left to do the work, but obviously could lack sufficient power and would stall.

In the gradability tests the problem of unconnected engines again showed itself. Here the weight transference and torque reaction could cause the front wheels to lose traction, placing the full drive load on the rear wheels which, though they had traction, lacked power. The rear engine would again stall.

Over-speeding of the unloaded engine, when one of its wheels lifted from the ground, was also a major failing. Speeds ▶

Twini-Moke with its covers removed. The rear engine's air intake is via the grille on the extreme left of the picture.

Two-Pack Workhorse

To the hills! A prototype Twini-Moke on test.

equivalent to 75mph could be reached by the freed wheel, with the result that when it regained grip massive shock loads were put on the driveline. Apropos of shock damage, the ratio of the bottom gear was too tall to allow the driver to crawl slowly over difficult ground without slipping the clutch. Hence the risk of damage by impact was increased.

Another problem was instability which could result under certain conditions when braking. This problem was not present on the mono-Moke, and was probably caused by the even distribution of weight of the two engines and the high polar moment of inertia of the Twini set-up.

But probably the most fundamental failing was lack of sufficient ground clearance. As tested by the US Army with the 12 inch wheels, the clearance was 8½ inches and the underpan would 'belly'. Such a failing is the death knell for a cross-country vehicle. The ground clearance was one bright spot; it was similar to the army's M151 jeep. Following these trials nothing more was heard from Uncle Sam.

SENTENCE: TRANSPORTATION

Ordinary Mokes were in production for four years in the UK from 1964. Then the taxman hammered one nail in the unfortunate machine's coffin by declaring it to be a car, with the result that extra purchase tax was attracted. About 90% of the British-built Mokes went as exports; only 1467 were home sales out of a total production of 14,518.

Then, in 1968, Moke production was transferred to the Leyland Motor Corporation of Australia, where it continues in a modified form to this day. Outwardly, the Aussie Moke appears similar to the original, though the wheels are now 13 inch, the track is wider and such refinements as windscreen extension visors have been added. Also, larger engines are now fitted, 998 and 1275cc, and front discs have been introduced. Other Australian developments include two-seater with a large pickup body, along with an upmarket version called the Californian which sports white spoked wheels and 'roo bars front and back. The Aussies have also produced a prototype single-engined 4WD Moke.

Amateurs have turned Mini into Twini, but no series production of Twini cars or Mokes has ever started. Of the BMC Twini-Mokes, an example is preserved in the BL Heritage Collection at Syon Park, Brentford, together with other Mini matters. Weekly production of the Mini has now been firmly overtaken by the Metro, but it's doubtful if anything can overtake the wealth of variation ideas spawned by good ol' Mini of 1959.

MINIOLOGY

These books all give information about the breed, and make interesting reading for Mini fans:
The Mini Story by Laurence Pomeroy, Temple Press 1964.
Mini by Rob Golding, Osprey Publishing 1979.
British Leyland—the Truth About the Cars by Jeff Daniels, Osprey Publishing 1980.
Amazing Mini by Peter Filby, Gentry Books 1981.

English Production Mokes

On 30th of January 1964 the first production Austin Mini Moke (A/AB1 513101) was dispatched from the Longbridge factory destined for Papua. Infact all the remainder of this initial batch of 10 (built January 1964) were destined there, or to Singapore. The first one for the home market (a Morris this time) was built in the second batch of 11 at the end of June 1964 and was sent to the Barton Motor Company in Plymouth. Within the first 50 Mokes built, at least 20 different countries were the recipients. In fact the British produced Moke was a bigger success around the world than in Britain itself (the climate might have had something to do with this!), for out of the 14,518 that were made during its four year production run, only around 10% (1,467) were for the home market.

As supplied, the Moke was very basic indeed. For the price of £335 you had a drivers seat, a hood, and just a single wiper blade, and you could have it any colour as long as it was Spruce Green. Optional extras included front and rear passenger seats, heater, side screens, sump guard, and Weathermaster tyres. In the last year of its production run a second wiper was added and the alternative colour of white was introduced. The basic Moke was the cheapest form of four wheeled transport on the market, helped by being classed as an utility commercial vehicle, and therefore not subject to the same purchase tax as cars. This made it appeal to many firms, and fleets were bought by the Devon Fire Brigade and Wimpey Construction amongst others. Unfortnately in 1967 the treasury changed its mind about it being a car and £78 was added to the price, destined for the governments coffers. This caused sales to drop off and production eventually ceased with the final Moke coming off the production line in October 1968.

The reviews on the following pages give some press reaction to the production Moke. The first is from the BMC produced commercial vehicle magazine Payload, so could hardly have been expected to be critical, whereas the second from Motoring Which tried to take the car rather too seriously. The others show the public's reaction to actually owning and driving a Mini Moke.

The demand for Minis soars, and here is the latest of the series to go into full production, the Mini Moke. Giving this one a final polish in C.A.B. No. 1 at Longbridge is Mr. Bert Smith, while Mr. Garnet Cox runs an expert eye over the interior fittings

A tough, practical little vehicle with a standard of 'over the rough' comfort belied by its appearance.

PAYLOAD TESTS AUSTIN mini-MOKE

$\boxed{30}$

FRESH air, we are told, is a blessing. We are also told that some blessings arrive disguised while others, more obvious, are not always appreciated. For the most part the *Payload* staff, deeply immersed in commercial vehicles during their working hours, are particularly partial to a breath of ozone, but after testing the remarkable Austin Mini-Moke over our standard 128-mile test route, we feel that here we must draw the line, and refute the opinions of hardened naturalists who declare that no amount of fresh air can do one any harm! There can be no doubt at all that the unusual, versatile Mini-Moke, buzzing around our test route like a worker bee, is probably the oddest and most certainly the airiest vehicle to be let out of the Austin commercial stables since their doors were first opened for a 15-h.p. van with solid rear tyres to splutter on to the market back in 1909.

Do not misunderstand us. The Moke as its name implies, is most definitely a practical proposition in a multitude of ways, and considerably more lively than its hairy cross-bred namesake.

The moke, *Encyclopædia Britannica* informs us, is a beast of burden, unimpressed by heavy weights, resistant to the elements, possessing sobriety, endurance, and sure-footedness with vigour and strength. The very same description can be tagged to Austin's mechanical work-horse announced in the autumn of 1964.

Although quite obviously in a class of its own, it is powered by the standard Mini power unit of 848 c.c. and the general layout of the engine/transmission is the same as that of the Mini saloon. Most chassis details are identical, too, which facilitates servicing and availability of replacement parts when needed.

The fully rust-proofed, unit construction steel body has a very low centre of gravity, which together with all-round independent rubber suspension ensures stability during spells of fast road work and a high standard of comfort over the roughest roads and even over virgin territory.

In this latter respect we were pleasantly surprised. Having thoroughly disguised ourselves in coats and scarves to confuse the elements, we discovered that even at high speeds—say around 60 m.p.h. (or over uneven country lanes at about the same speed)—the modest seating arrangement left us without a trace of those undesirable effects usually associated with long periods in the saddle. In fact, the Moke rode more evenly than many a private saloon, and a sample of the Editor's handwriting at a steady 50 m.p.h. showed a distinct improvement over the norm!

With front-wheel drive and the power unit mounted transversely over the front wheels, maximum traction is assured over difficult terrain, circumstances to which the Moke is ideally suited.

One of the great advantages of this sturdy vehicle which helps to increase its usefulness is the ease with which passengers and goods can enter and leave. However, we did agree that it was equally easy to make an involuntary exit from the Moke if a certain amount of care was not exercised by passengers during spates of unusually exuberant driving!

PAYLOAD TESTS

Uses to which the Moke can, and in many cases has already, been put include works transport for managers or maintenance men, farm work, airport ferrying and site survey work, as well as being used on the golf course as a caddy or by hotels as a beach wagon. There must surely be many other more varied uses but we leave it to the reader to decide just what these could be!

Taking a close look at the Moke we realized how well-constructed it was. Despite its stark but practical, angular appearance the vehicle is solid and rattle-free, likely to take a number of knocks without suffering unduly.

A stout bar, bolted to the front 'panel', affords some protection to the wings and grille alike. The box-like wings themselves rise well clear of the wheels, allowing clearance for mud which is certain to be collected in bad conditions.

The bonnet, held down by elastic rubber clips, can be removed completely in a moment to expose the whole engine assembly. In fact, almost all the mechanical parts of the Mini-Moke are embodied in this amazing unit. Only 18 inches of the total length of the vehicle are taken up by the engine, clutch, gears and differential, and drive-shafts. In the case of the Moke, the outstanding engineering achievement which made this possible allows for a very spacious carrying capacity.

Sitting in the driver's seat one realizes that the instrumentation is modest but adequate. The small frame containing the speedometer and steady-reading fuel gauge also mounts the ignition key and ignition warning light, choke, and light switches. For the sake of economy the standard single wiper blade is operated by a small motor. A second wiper and motor are available at extra cost for the passenger's side.

An interesting point about the seats is that the clip-on padded covering can be completely removed to leave the shaped metal frame available if the Moke is to be used for really tough work. The padding can then be retained in a clean

Usually market gardens are a maze of narrow paths and glass-houses. In such a situation the versatile Moke is an ideal vehicle. Less the rear seats, a very considerable space is available for carrying tools and produce.

Ferrying maintenance crews around airports or factories is a job the Moke is well adapted to. In these applications ease of entrance and exit are the important factors. Our test vehicle is pictured at Elmdon Airport prior to take-off of a B.E.A. prop-jet 802 Viscount 'John Hanning Speke'.

condition for more salubrious occasions. The additional front passenger seat is an extra £8 9s. 0d.

Rear seats are available at the extra cost of £16 18s. 0d. and are similarly constructed with cushioning clipped to sturdy metal frames.

The Moke is supplied with a vinyl-treated fabric tilt, supported by detachable tilt poles. This can be quickly erected by two persons and is not an impossible feat for one alone. The spare wheel, mounted in true military fashion on the rear of the vehicle, is quickly detachable by unfastening a single centre-bolt.

The braking system, as would be expected from a vehicle with a gross vehicle weight of only 17 cwt., is perfectly adequate. Lockheed hydraulic brakes with two leading shoes at the front and leading and trailing shoes on the rear, have a pressure-limiting valve to obviate rear wheel locking with excessive pedal pressure under lightly laden conditions.

The vehicle we tested was rather too new for our liking and did not, in some respects, give the performance that might have been expected. Whereas a maximum speed approaching 70 m.p.h. is possible, we recorded only 60 m.p.h. on the fastest section of our route. Nevertheless, an average speed of 43

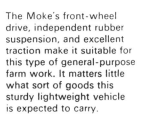

The Moke's front-wheel drive, independent rubber suspension, and excellent traction make it suitable for this type of general-purpose farm work. It matters little what sort of goods this sturdy lightweight vehicle is expected to carry.

Dwarfed beside an FJ K100 (5-ton) dropside truck on the site of a new building under construction at Longbridge, our test Moke was being used to transport surveyors and their delicate equipment over some very rough ground.

PAYLOAD TESTS continued

m.p.h. was maintained during Section Three of the test through the tortuous lanes among hills round Ross-on-Wye. Considering the fact that 60 m.p.h. was never exceeded, this surely is an expression of the Moke's unusual stability and road-holding capacity in such conditions.

At speed on the motorway the Moke did not appear as stable as its full-bodied cousin, the Mini-Van. This could have been due to the rather poorly distributed test load of lead shot or the parachute effect of the tilt. At lower speeds, 45–50 m.p.h., the ride was comfortable and reasonably pitch-free.

There can be little doubt that the Moke is an unusual vehicle with an almost unlimited potential in the commercial world. Tough over the rough and speedy on the road, the Moke has a standard of performance and fine weather comfort belied by its stark, purposeful but oddly attractive appearance.

ROAD TEST ANALYSIS—AUSTIN MINI-MOKE (G.V.W. AS TESTED, 17 CWT.)

Section	DESCRIPTION Route and Driving Conditions	Mileage	Time h. m. s.	Av. Speed m.p.h.	Fuel Pints	M.P.G.
1	Mainly A38 trunk road, from Longbridge to start of M50 (Ross Spur) Motorway; 35–40 m.p.h. cruising speed where possible and moderate throttle openings to result in over 30 m.p.h. average with reasonably fuel economy. Traffic light through Bromsgrove and Droitwich, heavy in Worcester. Rose Hill (1 in 8) climbed comfortably at 28 m.p.h. in third gear, and Severn Stoke (1 in 10 approx.) breasted easily at 28 m.p.h. in top. Engine cold and sluggish for first four miles or so	31·0	0 57 30	32·4	5·32	46·5
2	M50 Motorway to Ross Spur (including ¼-mile approach loop) at 40–45 m.p.h. cruising speed with light/medium throttle—a predominantly *uphill* section. Long 1 in 30 gradient reduced speed from 44 m.p.h. at bottom to 39 m.p.h. at summit in top gear on a steady throttle—a good performance from a fully laden vehicle with a (net) power-to-weight ratio of 29 b.h.p./ton as tested. Choice of final drive gearing (14·83 m.p.h. per 1,000 r.p.m. in top) gives good flexibility and excellent economy. Comfortable, pitch-free ride enabled observer to make legible notes at speeds around 45 m.p.h.; a surprising feature considering the nature of the vehicle under test	22·0	0 32 30	40·6	3·97	44·3
3	Unrelenting hard driving and stop-start operation through narrow, hilly, and congested streets of Ross-on-Wye and on subsequent uphill test section, culminating in 1 in 5 max. gradient of Bulls Hill. Here, engine torque, clutch and hand brake efficiency were assessed and a stop-and-restart accomplished smoothly in first gear only. Remainder of test via long, difficult 1 in 6 downhill section, followed by twisting, steeply undulating minor and major roads—via Symonds Yat—on full-throttle return to Ross Spur. Highest speed attained 60 m.p.h.	22·0	0 30 45	42·9	5·77	30·6
4	Return down M50 on part and wide throttle in generally downhill conditions to average nearly 45 m.p.h. overall. Cruising speed of 45–50 m.p.h. maintained, reaching an indicated maximum of 60 m.p.h. (4,050 r.p.m.) on downgrades. At high speeds the vehicle did not appear to be quite as stable as its full-bodied cousin, the Mini-Van. This could be due to a rather poorly distributed test load, or the parachute effect of the tilt cover. The creditable 'high-speed' 44·0 m.p.g. returned may be averaged against Section 2's figure of 44·3 m.p.g. to give a typical motorway figure of around **44 m.p.g.** laden at **42¼ m.p.h.** average speed	22·0	0 30 0	44·0	4·01	44·0
5	Final test—from M50 to Longbridge, mainly on A38 trunk road. Hard driving in generally uphill conditions in main road traffic of above-average density. Worcester was very congested. Open road cruising speed of 45–50 m.p.h. maintained where possible with liberal use of acceleration and indirect gears. This 34·2 m.p.g. can be balanced with Section 1's figure of 46·5 m.p.g. to result in a normal (laden) trunk road figure of **40¼ m.p.g.** at average running speeds around **34 m.p.h.**	31·0	0 52 15	35·6	7·25	34·2
	OVERALL: Harder than normal driving in above-average British road/traffic conditions	128·0	3 23 0	37·8	26·3	39·0

The Mini-Moke

It is pretty difficult to conceive of anything more basic than a basic Mini. But the Moke — looking business-like in War Department Green — manages it. At a basic list price of £413 it's nearly the cheapest new four-wheeled transport you can buy — and over £100 cheaper than the cheapest saloon Mini with a heater at £522.

Unfortunately, the basic price does not tell all. You don't get a front passenger seat (£9 extra), any back seats (£17 extra for two) — and without them it's illegal to go faster than 40 mph — or a heater (another £19 10s). If you're ever going to drive cross-country, you'll need a sump guard (£7). That little lot brings it up to £466, £50 more than a Fiat 500 at £417, a Mini-Van at £412 or a Mini Pick-up at £407.

And while BMC do give you a hood for your money, they don't give you side screens — you can get them from a BMC distributor for £30. But that brings the price to £496 — less than £30 under an ordinary heated Mini.

So your bank balance probably won't look much healthier if you buy a Moke instead of a Mini. What do you get instead?

What is a Moke like?

The Moke is a platform of roughly Mini dimensions with 4 seats on it and a hood over it. Apart from the driving controls and one windscreen wiper (the blade was too short — we replaced it with a longer one) there is no other furnishing whatsoever.

Mechanically the Moke is very similar to the Mini. So on the road you had the usual superb Mini handling and manoeuvrability, and sound brakes which are a little on the heavy side. The design has affected acceleration and top speed. They allow you to keep up with the traffic but little more — we found a top speed of 65 mph with the hood up and 61 mph with the hood down. You might find that a long journey would take you more time than you bargained for on a windy day, for head-winds slowed the Moke right down and indeed sometimes reduced you to driving in third gear on the level road to enable you to make any headway at all.

The primitive-looking seats, though small, were just about adequate. To make life bearable for the driver, though, we had to move the driver's seat $2\frac{1}{2}$ inches behind the position in which it was set when the car was delivered. But the thing that really would put you off long journeys in the Moke was your exposure to the elements. On a journey of any length the Moke gave you little more protection than a motorcycle — in the winter you would certainly have to dress up as though you were going to walk to the South Pole. The wind made the life of passengers in the back almost unbearable, and that of passengers in the front very Spartan. The hood made so much noise that it was almost certainly the noisiest vehicle we have ever driven. The pressure of the hood on the windscreen made it vibrate and so blurred the driver's vision. If it was wet, this vibration lifted the windscreen wiper off so that it did not wipe properly. And the wind brought in rain and mud thrown up from the road on to the inside of the windscreen, where you had to keep wiping it off.

You could keep the worst of the weather out, of course, by spending the £30 on a set of side screens, but driving surrounded by wet and dirty plastic, possibly difficult to see through, may well be little more pleasant. So we have to think about the

Moke mainly as a short journey, fine weather car.

Cross-country

To find out how useful the Moke would be as a general off-the-road workhorse, we gave it to a large agricultural estate in the North of England where conditions varied from long thickly-wooded valleys to high moors. People in a whole variety of jobs used the Moke, comparing it mainly with the fleet of 20 or so Land Rovers which were already in use.

There was no doubt that in conditions that suited it – hard, dry ground – the Moke was a much more pleasant car to drive than the Land Rover. Indeed on a fine day it was positively exhilarating to drive cross-country with the Moke's very precise steering control and manoeuvrability and on the whole very comfortable ride. The suspension coped surprisingly well with everything from mild bumps to the tremendous crashes you get when the car leaves the ground and comes down again – this is just as well as there is very little to prevent the passengers being thrown out of the car. But in conditions which didn't suit it, the Moke's cross-country performance was limited. The small wheels would drop into pot-holes and find it difficult to climb out again. The limited ground clearance of $6\frac{1}{2}$ inches could get you into trouble over deeply rutted cart tracks where the wheels would run in the track and the car would run up on its sump guard on the ridge in the middle.

We found it very difficult to keep the wheels pulling the car both in mud and on any sort of slope. Our agricultural users found that it would not go up slopes of more than 1 in 6 or 7, and we even had difficulty on comparatively gentle slopes of the order of 1 in 10, if they were at all slippery. In mud the wheels would spin very quickly and the car would dig itself in.

In fact with its small wheels and limited ground clearance the Moke would not go some places where a Land Rover would go in two-wheel drive, and it certainly could not go where the Land Rover could in four-wheel drive.

Another difficulty was that, because of its comparatively limited engine power, you had to use first and second gears a great deal in difficult conditions. And, because the Moke, like most BMC cars, did not have synchromesh on first gear, to keep going you sometimes needed to get into first gear regardless of noises of protest.

To carry anything very much in the Moke you would have to take out the back seats and there are then quite deep wells, though of course nothing is supplied to keep what you put in from bouncing out. A fairly major drawback in cross-country driving is that if there is any mud at all about it very soon gets thrown into the back, covering whatever is there in dirt.

It will be clear from all of this that most of the people who used the Moke for us did not think it was the answer to their workhorse problem. One or two, however, thought that it might well be useful. The shepherd found that his dogs loved riding about in it and positively barked for joy when it appeared. And he found that the back was a very good place for carrying dead sheep. The gamekeeper also approved of it – he liked the marvellous all-round visibility and thought that it would be very easy to conceal the Moke when he was out after motorised poachers. But the general verdict of our country testers was that the Moke might well be better for use in London where 'those fancy young women want to show off their knees' than in damp Derbyshire dales.

In town

So what about those fancy young women?

Certainly in fine weather the Moke is a marvellous town runabout. Highly manoeuvrable and easily parked, this is a fine car to drive in traffic – you can alternate basking in sunshine at the traffic lights with the feeling of standing in a stiff breeze on the sea front on a fine day. And of course it is a splendid mobile shopping basket. You can throw everything in and jump in and out yourself – but if you leave your mobile shopping basket full you will need to have some faith in your fellow human beings.

But again the picture changes as soon as it gets wet. Either you have to enclose yourself in plastic side screens in which case you lose all the Moke's advantages, or you have to be prepared to get wet. Driving through the back streets of Kensington in pouring rain in the Moke must rate, as an activity, very low on anyone's fun index.

The rest of the story

As far as surviving an accident goes, you are rather exposed in the Moke. You have virtually no protection from side impacts or if the Moke is rolled over – though it is virtually impossible to turn it over yourself. There is no padding or any sort of trimming in the vehicle and the front of the car contains numerous sharp edges and projections. We found that the seat belts did not give a great deal of confidence to our drivers because they still felt that they were not held securely sideways.

Over 5,000 miles the Moke averaged 33 mpg, but 3,000 of these miles were doing cross-country work where it managed only 27 mpg. It was extremely economical on oil.

As far as reliability goes, the fact that the Moke does not have any doors does at least mean that there is no possibility of the usual troubles of door handles not working and doors not being fitted properly. But sparse as the furnishings of the Moke are, they still developed various troubles — at the end of our test mileage the hood was beginning to split and its stitching was deteriorating, and the rear seat fixing strap had broken. It will come as no surprise to Mini owners to learn that the direction indicator lamp lens was broken too. The underbody was spotted with rust and the sides had been badly scratched where people had been getting in and out. Rust would eventually start in these scratches too. The Moke is quite a noisy car in any case and by the time our tests were finished the noise was amplified by the exhaust system being in contact with the sump guard.

Perhaps the most serious defects, though, were in the transmission system. We had to have a new drive shaft fitted under guarantee before 2,000 miles and the gear train was noisy by the end of our mileage. And, after our cross-country work, the engine stabiliser rubber bushes needed to be replaced.

Conclusions

You have to accept the fact that the Moke is not a car for long journeys. Once you have accepted that, the Moke is fine if it's fine, and not if it's not. Unfortunately, it suffers by not being as cheap as it looks. If you want a cheap country work-horse, the basic Renault 4 at £544 seems a much better bet. And in town, unless you are very optimistic about the weather, it does seem slightly eccentric to pay nearly as much as a Mini for a car which gives you such an unpleasant time in bad weather. If you can afford to keep a Moke tucked away to enjoy the fine weather when it comes, well and good. Otherwise we would think that you are likely to spend more time being miserable in the Moke than being happy. But we hope that BMC are making a fortune exporting it to San Francisco.

The Mini-Moke has long-wheelbase Mini layout, with a weather canopy which folds. An experimental 4-wheel drive model with twin engines front and rear has also been tested overseas.

mini-MOKE

Just announced in South Africa, the Mini-Moke is a low-cost runabout to appeal to farmers, contractors and holiday-makers.

ANNOUNCED as "the rugged runabout with a thousand uses", the BMC Mini-Moke is a utility version of the famous Mini, designed primarily for military use but well suited to a wide variety of utility purposes.

With its light weight, abundant power and front-wheel drive, it has applications on the farm and for surveyors, engineers and contractors.

It could also serve as a beach buggy or runabout, and it is welcome news that it has now been released in South Africa, priced at just over R1,000.

Based on the long-wheelbase (commercial)

NEW MODELS

Mini models, the Mini-Moke utilizes simplicity of design and rugged construction to achieve an indestructible character.

The four-wheel independent suspension gives it strong riding ability and considerable comfort for a vehicle of this kind, and the manufacturer claims that the Moke will perform well in mud, sand and snow.

Only 18 inches of the vehicle is taken up by the transverse engine, which has the same specifications and output as the unit used in the Mini passenger cars. For this reason it is capable of high performance if required and economy such as is seldom found in a working vehicle.

The instrument panel includes warning lights for ignition, low oil pressure, dirty oil filter element and headlamp high-beam.

It has seating for four people. Three of the seats are optional extras which can be removed quickly and easily to provide an unobstructed load space.

A collapsible canopy provides weather protection, and a sump protector and heavy-lug tyres are offered as options for off-the-road use.

SPECIFICATION

Engine: Water cooled, overhead valves, four-cylinder, counter-balanced crankshaft with three 1¾-in. main bearings. In unit with clutch, gearbox and final drive. Installed transversely at front of car. Bore 2·543 in. (64·588 mm.). Stroke 3·000 in. (76·2 mm.). Cubic capacity 60·96 cu. in. (998 c.c.). Compression ratio 8·0 to 1; maximum b.h.p. 39 at 5,000 r.p.m. Maximum torque 52·5 lb./ft. at 2,700 r.p.m.

Fuel System: Single S.U. Carburettor, type HS2; S.U. electrical fuel pump; air cleaner with paper element; petrol tank capacity 6¼ gallons (28 litres); fuel filters in pump and fuel tank.

Ignition System: 12-volt coil, and distributor with automatic and vacuum control. Suitable for operation with commercial grade fuel.

Cooling System: Pressurized radiator with pump, cowled fan, and thermostat. Capacity approximately 5⅛ pints (3 litres).

Transmission: Clutch: 7⅛-in. (0·18 m.) diameter, hydraulic operation by pendant pedal. **Gearbox:** Four speeds and reverse with synchromesh on second, third, and top; in unit with engine and final drive; central floor-change speed lever. **Final Drive:** To front wheels via helical spur gears and open shafts with universal joints; drive casing in unit with engine and gearbox.

Steering: Rack and pinion; 2⅓ turns lock to lock. Turning circle 31 ft. (9·4 m.).

Suspension: Front (includes final drive): Independent with levers of unequal length. Swivel axle mounted on ball joints. Rubber springs and telescopic shock absorbers mounted above top levers. **Rear:** Independent trailing tubular levers with rubber springs and telescopic shock absorbers.

Brakes: Foot: All four wheels, hydraulically operated by pendant pedal with two leading shoes at front and trailing shoes at rear. **Hand:** Central pull-up lever which operates on rear wheels. In order to achieve positive braking a pressure limiting valve is introduced between the master cylinder and rear brakes.

Road Wheels: Pressed steel, 4-stud fixing; 5·20-10 four-ply tubeless tyres. Optional for cross country, 5·20-10 Weathermaster tyres, with tubes.

Instruments: Speedometer with petrol gauge, including warning lights to show dynamo not charging and headlamp high-beam position, low oil pressure and dirty oil filter element. The various switches, including combined ignition/starter switch, are mounted on a panel in the centre of the fascia.

Bodywork: Pressed-steel unitary construction, open-type body with vinyl-treated fabric tilt cover, supported by detachable folding tilt tubes. Fabricated pressed-steel sub-frames, detachable from the body, provide mounting at front for power pack/front-wheel-drive assembly, and for trailing arm suspension elements at the rear. Seat of pressed-steel construction is detachable, and has a limited range of fore and aft adjustment. Hinged bonnet top is also detachable. Windscreen can be folded down or removed completely. Strong tubular bumpers front and rear.

Leading dimensions and data:
Wheelbase .. .. 80 5/32 in. (2·03 m.)
Length (overall) .. 120 in. (3·05 m.)
Width (overall) .. 51½ in. (1·36 m.)
Height (hood to ground) .. .. 56 in. (1·42 m.)
Height (screen to ground) .. .. 51¾ in. (1·31 m.)
Ground clearance .. 6⅜ in. (16·2 cm.)

Optional Extras: Sump protector, Dunlop Weathermaster tyres front and rear, front and rear grab handles, windscreen washer (manually operated), laminated windscreen, passenger seat, two rear seats. ●

Gear ratios:	Gearbox	Overall	Final Drive	Road speeds at 1,000 r.p.m.
Reverse	3·628 : 1	13·659 : 1	—	
1st	3·628 : 1	13·659 : 1	—	4·086 m.p.h.
2nd	2·172 : 1	8·178 : 1	—	6·825 m.p.h.
3rd	1·412 : 1	5·316 : 1	—	10·499 m.p.h.
Top	1·000 : 1	3·765 : 1	3·765 : 1 (17/64)	14·824 m.p.h.

Spotlight on That Preposterous Mini-Moke

Driving it is a ball—mainly because it's so uncomfortable, so maneuverable, and so wacky

By Alex Markovich

HA, HA, HA, HO, WHEE-HOO . . . I was just . . . just driving the Austin Mini-Moke, and I'm . . . hee, hee . . . still hysterical.

I started snickering when I first saw it; it's basically the Mini sedan so popular in England, but it looks like an un-Sanforized Jeep that's been stripped by vandals. No side windows. No doors. You just sit there, perched on the brink of disaster, your knees scrunched up because the seat doesn't adjust.

The tiny 34-hp chugger takes 24 long seconds to reach 60 mph. The drive train is noisy. The ride is harsh. And then (the horror, the horror) it begins to rain, and

IT'S GREAT FOR CITY TRAFFIC; you can slip into tiny openings that other drivers don't even notice. And pretty soon, you get used to people pointing and snickering—if you're tough. But heaven help you if it rains

passing cars get you all wet and everyone seems to be laughing at you even more than usual.

And yet I enjoyed this car perhaps more than any other I've driven. The cornering is better than that of most all-out sports cars. Just keep the throttle wide open through the turn and the front-wheel drive pulls you through. If you go in too fast, the front end starts to plow; so you let up on the gas and the rear end lightens, putting you in a perfectly controlled drift. It's a car that's easy to steer with the throttle.

The rack-and-pinion steering is sensitive, vibration-free and, at 2⅓ turns lock-to-lock, extremely fast. Directional stability is excellent, even in strong crosswinds.

The brakes are more than adequate, and they don't grab. The four-speed gearbox, however, is archaic, with the top three gears poorly synchronized and first not synchronized at all. But it's easy to double-clutch into first while rolling. The clutch is ferocious. When you're engaging it, nothing happens until the last couple of inches of travel, and then suddenly the seat back is bending your spine.

The dashboard holds only a speedometer, a gas gage, a few warning lights, a choke knob and a headlight toggle. And the gas gage isn't really necessary; the tank, built into the left frame member, has a mammoth filler cap that allows visual inspection of the gas level. A lone, spindly windshield wiper bravely defies the elements. It has to be started and parked manually.

And then there's a rag top that drums at speed, but folds easily and unbolts entirely in seconds.

Okay, but what's the car good for? I was afraid you'd ask that. I don't know, really, but on the other hand I can think of a dozen things. Maybe as a golf-course buggy or hotel runabout. Or (at $1270) as basic transportation in warm, dry climates. One thing it's definitely good for is fun. And it's good for shy, introverted people who have trouble making friends. When you drive a Mini-Moke, strangers stop and talk to you. The car is lovable because it looks so small and helpless, and strangers tend to transfer some of this affection to you, too.

"How do you like your roller skate?" they ask good-naturedly.

"Do they allow it on the streets?" one Volkswagen owner asked incredulously.

"Yes," I answered. "Do they allow yours?" End of conversation.

But there are also the less pleasant types, like the convertible-full of Beatle-mopped urchins who suggested I give the car back to the amusement park—but in less polite words.

Maybe the Mini-Moke is for extroverts after all. ★ ★ ★

BADLY LOCATED DIRECTIONAL STALK keeps getting in the way of the knee; may suffer a 'sad fate (above). Center speedo works for right or left-hand drive

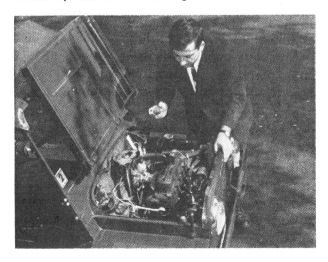

TINY 34-HP, 51.7-CU.-IN. FOUR is mounted transversely, from east to west. Though it's a tight fit, frequently serviced parts are accessible

The Golf Caddy of the 70's

The already familiar Mini-Moke has now been put on sale to the public. It is built as a go-anywhere, do-anything, maid of all work

IT was two years ago when we first reviewed the little cross-country derivative of the Mini-Minor. Since that time there have been a number of refinements and variations to the original design. The twin-engined Moke, which did such sterling service carrying feed to cattle and sheep in the bad winter of 1962/3, was the first development. On the single-engined version models have been made for both the British and American armies to test.

Now a standard version has been announced for sale to the general public. Although basically a commercial vehicle, it is likely to prove popular doing such odd jobs as a beach runabout, holiday camp taxi, golf course caddy vehicle, and a maid-of-all-work for factory transport.

As a hardy and economical little runabout the Moke seems difficult to fault. The robust body will stand any amount of rough work and will carry an extraordinary amount of equipment. The single adjustable seat is detachable, as are the bonnet top and windscreen. The latter can also be folded flat. To give all-weather protection a vinyl-treated fabric tilt cover is supplied as standard equipment. Strong tubular bumpers are fitted front and rear.

The engine is the standard B.M.C. 'A'-series 848 c.c. unit which is fitted in its low compression form. It develops a maximum of 34 b.h.p. at 5,500 r.p.m. and has a maximum torque of 44 lb. ft. at 2,900 r.p.m.

One feature that is likely to have great appeal to users is the sort of economy that goes with a Mini. On the road the Moke returns the same m.p.g. figures as does the saloon. This, coupled with the traditional Mini roadholding and handling characteristics, makes the Moke one of the most versatile and safe cross-country vehicles on the market today.

In performance too the Moke is no slouch. With a top speed of 70 m.p.h. it returns approximately the same acceleration times as the Mini-Vans. The crossways engine layout of course means that the engine, clutch, gearbox, and differential only require 18 inches of the total length. This leaves 8 ft. 6 in. for passengers and load.

The instruments are housed in a small panel set in the middle of the fascia. Contained in the speedometer are also a mileage recorder and warning lights for ignition, oil pressure, and headlamps high-beam. Also on the panel is a warning light which tells you when the oil filter needs changing. In addition, the panel houses the choke and light switches.

Ever since the prototype Moke was first announced there has been considerable interest in its future development. Certainly at £406 there should be no lack of demand for it.

Above, test car waits impatiently on gas line at Bridgehampton Raceway behind 160-mph Lola. Below, Moke can hide behind low walls.

THE MINI MOKE IS NO JOKE

And it's not a jeep, not a beach buggy, not England's answer to the Volkswagen. BMC's skateboard with an engine is simply one of the most fun-packed 'cars' ever built

THIS WRITER road tests all kinds of cars. Some are built for performance—for acceleration and the thrill of power (see Olds 4-4-2 road test on opposite page). Other cars are built for utility—to provide good, economical transportation. But one car is built for neither speed nor economy. It's the Mini Moke manufactured by British Motors Corporation. The Moke is meant for one thing—fun!

Statement of opinion: This writer has never in his career of testing all kinds and types of automobiles had more fun with any one car than he had driving the Mini Moke for two short weeks. The Moke (it's British slang

BY FRED MACKERODT

Front end look of the Moke is really distinctive, but the side view verges on the unbelievable.

With rubber 'springs,' Moke handles like a car three times its weight. Tires are very small.

for donkey) is just about the most different and enjoyable car he's ever driven. During the test, he found himself being asked for a ride by a foot-sore campaigning U.S. Congressman (who said he wished he had a Moke to use as a campaign car because the Moke attracted more attention than 5000 banners and loudspeakers), and also being picked up on two different occasions by attractive females!

The Mini Moke was first shown by BMC in 1964. Although it looks like some kind of military vehicle, it has no military applications. (American and British Generals did have a look at the Moke when it was first introduced, but since they had no use for a fun car, they forgot about it.)

It also resembles a beach buggy, but it can't make it on sand. We, in fact, tried negotiating the dunes with our test car and found ourselves up to the axles in the nitty-gritty stuff before we could move five feet. The BMC public relations honcho, Peter Smith, had warned us not to drive the car on sand, but we just have to try everything at least once. Getting bogged down was no big problem, however. A bunch of muscle boys simply picked up the 1200-pound skateboard (with yours truly contributing verbal support) and put it back on the black-top.

The Moke looks funny at first glance, but you can't tell how funny it really is until you crawl behind the seat, fasten the seat belt, and move out in traffic. The sensation is like nothing else imaginable. The car attracts more attention than Brigitte Bardot in Boys Town. Because your whole body—except for the lower part of your legs and feet—is right out in the open, the people staring at the car make you feel self-conscious—almost as if you were sitting there in your birthday suit. But you get used to it after a while, and you begin to enjoy the notoriety (if this happens to be your hang-up, too).

Getting down to nuts and bolts, the Moke is a direct descendant of the Austin Mini, which in "Cooper S" trim won the Monte Carlo rally an unprecedented four times in a row. The Cooper version has also been putting down the supercars in Class C sedan racing.

The Moke has basically the same transverse-mounted

Moke handles like a baby carriage at lower speeds. Above, we lead-foot under Lowenbrau overpass at 'Bridge' during performance tests.

Moke even attracts the attention of professional race mechanics.

Engine is transverse-mounted four-cylinder with whopping 883 cc's and 37 hp. Speedo reads to 90 but real top end is 70 mph.

engine and front-wheel drive as the Cooper, but the Moke has only 883 cc's as against the Cooper's 1275 cc's (37 hp compared to 75 hp).

While the Moke has much less power than its racing brother, the heritage is still apparent. The engine is smooth and winds nicely, pulling the car up to a 70-mph top end. We drove the car flat-out on our way to the Can-Am races at Bridgehampton Raceway for almost 100 miles without a hitch. The speed isn't apparent until you look down at the pavement whizzing by about two feet from your posterior.

The car isn't really meant for super-speedway driving, and it will keep up with traffic in any city-town situation. Or, it can be put into Cooper S trim for a basket of sheckles. This will boost the top speed to almost 100 mph, and we'd hate to think what it would be like to look down at the pavement going at that speed! We'll stick with 883 cc's. The car accelerates well at all speeds under 60 mph—and that's good enough for us. It also racks up 40 mpg.

Handling is where the Moke really shows its racing lineage. The Hydrolastic (rubber) suspension provides a sensational ride, considering that the car weighs only 1200 pounds. These rubber "springs" give the car a solid, sure feeling and make the Moke handle like a car with three times the weight. Cornering is superb. With the front-drive pulling the Moke around the hairiest corner with all 883 cc's screaming, the suspension keeps the car flat with very little lean or mush. You have to drive it to believe it.

Our test car proved to be a very versatile vehicle at times. Once we had to move a double bed which wouldn't fit into a big Chrysler station wagon, so we merely laid it over the rear section of the Moke with both ends of the bed protruding about two feet on each side. This isn't exactly legal, but it's just one practical application of the Moke's "open-ness."

The car is delivered with a hand-folded top, but without side curtains. We've seen some Mokes with custom-fitted bodies and some with custom side curtains, but as far as we know BMC doesn't offer these accessories as options. For this reason, the *(Continued on Page 34)*

MINI MOKE
CONTINUED FROM PAGE 33

Moke in stock condition is really a fair-weather car.

So if you want to have a ball and experience a really way-out driving sensation, take a ride in a Moke. If you live near Frank Sinatra, Elke Sommer, or Trini Lopez, bug them for a ride. They all own Mokes.

And if you're interested in buying a Moke, you can purchase all 1200 pounds of it for a buck a pound. But you'd better high-tail it down to your local BMC dealer plenty quick because the Mini Moke will no longer be available in '68. It will no longer be available to *Americans*, that is. BMC has no intention of discontinuing the car. The Moke is being banned from the Land of the Free and the Home of the Brave because of safety legislation. The car just can't hope to meet the rigid government standards set for '68. And the government, like most bureaucratic institutions and Mrs. Murphy's corset, isn't just likely to flex an inch from the prescribed course it laid out for all car makers, both domestic and foreign.

Although the Mini Moke handles better, brakes better and is a safer car to operate than a lot of accepted American garbage pails we've had the extreme displeasure of driving, it will be condemned because of its "open-ness," along with a host of other nit-pickin's. There just isn't enough sheet metal around the driver to *protect* him in case of an accident, says Washington. That's the big safety word today—"protect." Big Brother Safety in Washington has shifted his stance from the very laudable Nader-type principle of pressuring the auto makers into remedying certain unsafe conditions and thus *preventing* accidents, to that of *protecting* people, who don't really want to be protected. The safety people have shifted their tactics from prevention to protection, and this, we don't think, is the American way—but it is the way the safety ball is currently snowballing.

For instance: Early in the safety movement Washington required that automobiles be equipped with safety seat belts—a preventive measure. But they never required that the safety belts be worn—a protective measure. However, when they got around to motorcycles, Washington asked the state governments to require the *wearing* of crash helmets—a protective measure. Most cyclists now don't have the right to choose whether or not they want to wear a helmet—they have to, or else take a chance on getting a ticket.

Well that's the way it is.

THE donkey has always been an important animal in East Africa. From the burning semi-desert country of the Northern Province of Kenya to the cold wet mountains of Southern Tanzania you will find him plodding his way along roads, game trails, and mountain tracks. He has many attributes which make him particularly suitable to the needs of Africa. A donkey can go almost anywhere, his running costs in the shape of food are low for he can frequently live off the country. He is generally good tempered and willing and is the pride of his owner.

But the ways of Africa change perhaps even more rapidly than those of the rest of the world, and a new breed of donkey has recently arrived in Africa. They don't quite call it a donkey but it possesses all the attributes I have mentioned above plus a few more. I refer to the Mini-Moke.

Africa can be called the home of the cross-country vehicle for, let us be honest, many people from outside the continent when they were driving here would not know when they were on a road or when they were going across country.

Small fingers of tarmacadam roads are reaching out across the vast spaces of East Africa but the majority of our roads are still those labelled 'all weather'. This means that in dry weather you can go from A to B on a rough, corrugated surface rather like a poor English farm track. Because of the distance from *A* to *B* this must be covered at a fairly high average speed, which entails considerable discomfort and a diet consisting chiefly of dust. On the other hand, when it rains it rains. The dust, of course, disappears, to be replaced by some of the softest, most clinging mud you will find anywhere.

Hazards there are in plenty. When people in the more developed countries talk of drifts they may be referring to large piles of snow, or if motor-racing enthusiasts perhaps of Jack Brabham's coming fast through a corner. Drifts to the East African motorist are something different again. We have numerous watercourses in this country which for long periods are merely dry river-beds. Money is scarce and when these cross a road, which is pretty frequently, the road drops down to the river-bed, crosses it, and climbs the farther bank. This is an East African drift. When the rains come, however, these dry river-beds can become roaring torrents which can only be negotiated with care. If the water is too deep, one has to wait on the bank with patience for the flash floods to subside.

Corrugations are, as the name implies, a road surface which resembles a gigantic piece of corrugated iron. The effect on the suspension of driving over these for long periods at high speeds can be imagined. Not only is it wearing on the vehicles but the effect on the driver is the equivalent of many miles of horse riding. I am told it is very good for the figure.

The suspension on the Mini-Moke is tough as befits its name but it is also surprisingly comfortable and does take away a lot of the strain of the continual bashing for the driver.

The two requisites of a safari vehicle, not to mention the driver, must therefore be toughness and the ability to cope with difficult road conditions. We have had no shortage of good cross-country vehicles in the past but they have all suffered from two big disadvantages. They are comparatively large and fairly expensive.

Low price and running costs

Much of our safari work is done by government officials who usually travel on their own or at times with one assistant. It is nearly always wasteful to fuel, therefore, to use a large vehicle. In addition, Tanzania is a developing country which needs to use every penny of its resources to the best advantage. The Mini-Moke costs about half the initial price of its larger competitors and is far more economical in its running costs.

The number of uses to which people put the Mini-Moke is multiplying daily. The forestry officer driving his way through the narrow, twisting forest trails closely lined with his charges uses a Mini-Moke. The field officer on a large sisal estate and the housewife who wants to run into town for a loaf of bread all swear by their successor to the donkey.

Perhaps the real proof of the popularity of the Moke however, comes from BMC itself who in their production schedules did not allow for anything like the demand which has developed, giving prospective purchasers in East Africa many months on a waiting list.

THE TAILLESS DONKEY
by NORMAN TAYLOR

Above. If you are too lazy to walk across that stretch of hot white sand for your swim, the Moke can help you there, too

Below, centre. The Moke arouses interest as it drives through a fishing village

Below, right. Happy circle

McMOKE

ONCE UPON A TIME AN UNGAINLY LITTLE MOKE AND A SLEEK black Jaguar met at the foot of a snow-clad hill in Scotland (Cairngorm to be exact).

'Let's see who can get our passengers onto the ski slopes first,' suggested the Moke who, though lacking grace, was extremely adventurous and enjoyed a challenge.

The Jaguar snarled its derision, leapt forward, and within moments had disappeared round the next bend.

Purposefully the Moke followed it up the tortuous, treacherous road. The story ended at a height of 2,400 feet when the Moke eventually found the Jaguar embedded and impotent in a snowdrift. All of which confirms that not only does pride precede a fall and all that glisters is not gold but that 'Never go without a tow rope' is almost a duty as far as Moke owners are concerned. Since that initial occasion on Cairngorm we have towed two cars out of snow and one out of deep mud.

We bought our Mini-Moke when the family started ski-ing and our sports cars proved useless as beasts of burden. Like its four-legged namesake the Moke is ugly, a continual butt for humour, and an indefatigable worker.

Fitted with all-weather tyres it shoves its blunt little nose straight into blizzards and plods on through snow and over ice. Frequently the Scottish ski resort car parks are too tightly packed for mental comfort but here again the Moke is at an advantage as no one sliding into it could possibly ruin its looks!

As it arrived ex-factory the Moke was undeniably cold. Riding around in it, particularly in north Scotland, was as snug as living in a tent with only the central pole and guy ropes. Very early on, therefore, we had sidescreens of Lionide tailored, with a rear window of very pliable Perspex. The screens are easily studded on and off, and cost just under £15. For another £8 we fitted a heater and although, inevitably, some of the warmth disappears through the junction of canvas and metal the temperature within has been raised well above the original freezing point.

For non-ski-ing outings I put carpeting on the floors (back and front) and a thick wadding of rugs on the back seat. With the roof down and screens off the Moke thus becomes, on really good summer days, an ideal tourer. Short in the body though it is (nine and a half feet from withers to croup) it transports a family of five, sporting or picnic equipment, and a Labrador.

Apart from doing its stint as family saloon and ski bus, one week-end it became a Territorial Army ambulance and carried a parachutist with a badly fractured shoulder through marshland, then up and over some sizeable sand-dunes. It is also used by our varied-interests family as pop group transporter, timber haulage wagon, boulder shifter, and caravan—the latter when we drove the two younger children 200 miles through the night. Well padded with quilts and cushions the broad back seat made, they assure us, a most comfortable bed.

For town use it has two advantages: it has so much space for parcels and it takes so little space for parking. All that's needed for the latter is a gap one foot longer than the Moke itself and, once you've nosed it in, a man to lift the rear end into alignment. Its two town disadvantages will, presumably, lessen with time. One is the laughter it evokes, the other is the time it takes to answer the inevitable questions about performance, petrol consumption, and so on. Our reply, incidentally, is that my husband does a 48-mile journey by Moke each morning. This takes him 55 minutes and one gallon of petrol, and he avers that it corners better than most conventional cars.

Of the eight friends who have used our Moke, four, for an intesting cross-section of reasons, have subsequently bought or ordered one. A family of camping enthusiasts are about to set out for France in theirs, one friend plans to drive overland to Rhodesia in his. Another is to use it for a milk round, and the fourth, who has very little money but needs a vehicle, considered it a best buy.

One of the youngest Moke drivers, however, must be our seven-year-old son. After two lessons he could control the Moke sufficiently to drive on his own and now, as our driveway is virtually impassable for normal cars, he is dispatched to meet guests at the gate and bring them up to the house by Moke. He also drives it up moorland tracks, over burns and down heather-filled gullies in an attempt to earn his keep as a gatherer of firewood—and so far the Moke's only trouble has been a displaced exhaust pipe!

Admirable though it is, I had not intended using the Moke for my own holiday transport this autumn. In early spring, however, we acquired a lamb which has since become an unbelievably cumbersome ram and which becomes obstreperous when left for any length of time. But perhaps after all the sight of me touring the West Coast in a Mini-Moke with one child, one Labrador, and one sheep won't cause any more surprise than when we drove out for the first time in the town's first Moke.

Mini Moke

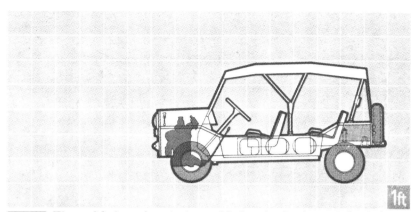

William Brown is a 25-year-old mechanical fitter who lives in Sheffield with his wife and one-year-old son.

In my job I have had a lot of experience of looking after Austin Champs and Land-Rovers but those I handled gave so much trouble that I decided we would be better off with a Moke. Apart from wheel bearing trouble, which we had fixed in Paris on a holiday, it has been completely reliable.

The Moke is marvellous in good weather, but the soft top wouldn't stand up to strong winds, and when the baby arrived I built a detachable aluminium hardtop with sliding windows and opening rear door. I also fitted a rear seat, heater and a tow bar. It's now very comfortable and still a lot of fun.

PRICE
on the road	**£501**	**6s**
basic	335	0
p.t.	78	9
road tax	17	10
average delivery	8	0
seat belts	6	0
wing mirrors	2	10
side screens	23	0
back seats (2)	17	4
heater	13	13

COST
per week	**£5**	**6s**
running	2	17
depreciation	2	9

AA The Moke is baby brother to the Champ and Gypsy general purpose vehicles, and though it does not have four-wheel drive it is quite happy to leave the road and cross not-too-rough country. In the right weather it's a great fun car. In a winter rainstorm it can be decidedly depressing.

Mechanically, of course, the Moke is mainly Mini. Lively performance is accompanied by a sporty exhaust note and the top speed is an exhilarating 72 mph.

The brisk acceleration is partly due to the excellent transmission. The lightweight clutch bites with a snap and will re-start the Moke on a 1-in-3 hill or whip it smartly away on the level. Rapid gearchanges are made easy by powerful synchromesh on the upper ratios and by the poker-straight gear lever which is much easier to handle than the Mini saloon's bent lever.

Suspension is by the pre-Hydrolastic system of rubber cones, and the satisfactorily firm ride on good surfaces becomes choppy on cross-country passages.

Nevertheless, the tough little Moke claws its way over rocky gullies impressively. On decent surfaces the handling is typically Mini and the Moke can be flung into corners with gusto.

In its spectacular lack of creature comforts the basic Moke rivals the original Jeep. It comes with one seat—a metal bucket with clip-on foam cushions—and a soft top. Side-screens, heater, and passenger seats are extras. Driving in the wet without side-screens is like sitting in a shower cubicle with your clothes on. BMC thoughtfully provide drain holes in the floor, and a rubber mat brings the nearest hint of luxury to the front compartment.

The driver sits close to the upright wheel, with bent legs working the small pedals. Exterior mirrors are standard and vision is still good with the soft top on, until the flexible rear windows silt up.

One of the wide side-members of the Moke holds the petrol tank; the battery is in the other; and there are storage boxes reached by quick-release panels. The spare wheel is mounted externally at the back, and in its four-seater form the Moke has room for a small amount of luggage behind the back seats.

SPECIFICATION
Engine—848cc, 4 cylinders, producing 34 bhp at 5,500 rpm, oil change every 6,000 miles
Transmission—4 forward gears, no synchromesh on first
Suspension—independent all round
Brakes—drums all round
Tyres—5.20 x 10
Fuel—4 star, 6¼ gal
Body—open type, doorless.
Colours: snoberry white, spruce green

'Oh, NO!'

Australian Production Mokes

In 1968, production of Mokes stopped in England, and was transferred entirely to Australia. The Australians had been making Mokes since 1966 on a small scale, with some improvements to make them better suited to local conditions, including a larger 1100cc engine.

These improvements were gradually introduced to the full scale production, along with other developments, but the underlying vehicle was still the same. The first significant alteration was to change to 13 inch wheels (from 10 inches on the English Model). This improved the ground clearance but necessitated the widening of the vehicle at the front and rear, and also did no good at all to the turning circle. Then came the larger engines and a pickup truck (see photo below) leading onto the upmarket Californian, sold alongside the standard model, and incorporating roo bars front and rear, roll bar, white sporty wheels and comfortable seats. Larger engine sizes also became available including the sprightly 1275cc.

All seemed destined for a long future, the Australian public liked it, and they were still sold in numbers to the hotter climates around the world, some even found their way back into the UK thanks to the enthusiasts at RunaMoke. But again time ran out for the Moke, the factory was becoming uneconomic, and the assembly line was closed down in 1982 to make way for the more profitable assembly of Peugeot cars. At the time, complete, but not yet ready for production was a 4x4 Moke, based on a single engine this time, and also the drawings for a rear engined sports car consisting of a fibre glass top placed over the original Moke bodywork.

The following pages describe the vehicles of this period and include an interesting comparison with the Citroën Mehari, a vehicle similar to the Moke, but derived from the 2CV. The article from the Australian magazine Bushdriver *We Poke The Moke* gives some useful pointers as to the reason for the eventual closure of the factory in Australia.

car Review

MOKE

Is it an all purpose vehicle for the countryman or a runabout for the "IN" crowd?

While the British Motor Corporation's all-purpose Mini-Moke will never win any prizes for looks, its versatility and ruggedness in hard going will impress those looking for a dependable 'work-horse' in the light commercial range.

As the prefix to the Moke's name suggests, it is a variant of the famed Mini-Minor sedan, and in its own particular field as a sort of "jack of all trades" vehicle, it may well prove similarly succesful.

The engine under the bonnet — which looks something like a sawn-off version of the old Willys jeep — is basically the same as used in the Mini deluxe.

Its capabilities insofar as acceleration and maximum speed are concerned, are limited by the fitting of lower ratio gearing in the four-speed box. This is a logical move in keeping with the Moke's prime purpose as a load carrier, and the need for hill climbing ability and flexibility in rough country off the beaten track.

To test the Moke in this role I ran it across several ploughed paddocks, well churned up and softened by recent heavy rains. Although unladen, which reduced the likelihood of bogging down in the mire, it surged through all obstacles quite effortlessly.

The fact that it is a front wheel drive vehicle also helped it conquer the course, a point demonstrated two or three times when the Moke seemed to "claw" its way out of especially soft patches. But even if I had become bogged past the point of getting out under power, there would have been no need to summon the nearest tow truck.

The Moke weighs only 1255 lb, at kerbside and can be lifted comfortably by four men.

BMC describe the Moke as having a "buckboard body". The term, conjuring up as it does the impression of an almost rigid, spine jarring ride for passengers, is slightly unkind to the Moke.

Certainly its only two seats are unsprung and have little cushioning, but on a reasonable road surface the Moke skips along without too much fuss or clatter and passengers should not find the ride too bad at all.

Like the work-horse it imitates, the Moke boasts few creature comforts. In its basic form it is open to the weather on all sides with no protection for passengers other than a vinyl-treated fabric hood; a heater is an unheard-of luxury.

For the less hardy, wind deflectors and a full set of four side-curtains can be fitted for an extra $30 — and if these had not been fitted to the test vehicle, one or two tropical downpours I encountered during the run would have come close to setting the interior awash.

As it was, they afforded a reasonable degree of protection, despite a number of spots where the driving rain managed to force its way through.

Behind the two front seats, the all steel welded body — which inside much resembles a small scale wartime amphibious vehicle — offers 41 inches of loading space to the rear. Pannier boxes are incorporated along each side of the body, between front and rear wheels. These house the battery, a 6¼ gallon fuel tank, the tool kit and jack.

For protection in rough going, the underside of the body is shielded by a substantial sheet steel sump guard attached to the sub-frame. Other body fittings include lap-type safety belts. Safety chains are available as an optional extra, for fitting on the outside of the body, beside the driver and passenger seats.

The heavy bar tread tyres fitted as standard undoubtedly help to stiffen the general ride of the Moke, but they do not affect the handling.

Steering is quite light and manoeuvrability good. In fact it is so good that the Moke, with its turning circle of only 30 ft., can be squashed into a kerbside space which, at first glance, looks an impossible proposition.

Overall it is an easy vehicle to drive, although the test unit needed a gentle foot on the clutch when moving away in first gear, as the drive take-up was inclined to be a bit fierce.

Braking, too, was slightly on the "sudden" side, although I would not call this a fault. Altogether, considering there was nothing at all elaborate about the Moke's solely drum-type braking system, its ability to stop in a stride was remarkable.

Under load carrying or low gear conditions, BMC

says the Moke will give approximately 26 mpg fuel consumption. During the test, in an unladen condition, the Moke returned an overall consumption of 40 mpg, with 36 mpg easily obtained cruising at a steady 50 mph. Hard driving was the order of the day, and these figures possibly could have been bettered by taking things more easily.

Instruments in the interior are at a minimum, comprising speedo, fuel gauge and generator charge meter, and warning lights for oil pressure and headlight "high-beam" operation.

To sum up: A multi-purpose newcomer of unusual design and potential.

SPECIFICATIONS

Engine: Four cylinders, bore 64.58, stroke 76.2, capacity 998 cc, compression ratio 8.3 : 1, power 38 bhp at 5250 rpm, 52 ft lb of torque at 2700 rpm.

Transmission: Four forward speeds with synchromesh on the top three ratios.

Suspension: All independent. Front — rubber springs and telescopic shock absorbers. Rear — trailing arms, rubber springs and telescopic shock absorbers.

Brakes: Hydraulic drum.

Steering: Rack and pinion. 30 ft turning circle.

Dimensions: Wheelbase, 6 ft 8 ins; Track, front —3 ft 11¾ ins. Rear— 3 ft 10⅞ ins; Length, 10 ft 6½ ins; Width 4 ft 3½ ins; Height 4 ft 10 ins. Ground clearance, 6⅛ ins; Weight, 1867 lbs.

MOTORING — MOKE STYLE

BLMC do build a modern sports car with independent suspension, transverse almost, mid-engine, and way-out body styling. Its name: Moke, of course.

IS the BLMC Moke the only modern sports car BLMC build? Is it a go-anywhere Land-Rover type vehicle? Is it a simple city runabout for small delivery services? Is it a practical piece of transportation? Or is it just a fun car for use during the summer? It is all these and more. During my week with the new 1100 Moke Mk. II, I left more expensive and glamorous road test cars at home to enjoy motoring Moke style.

It is taken for granted as a fun machine, but unlike many such vehicles, you don't begin to tire of the Moke after a couple of hours behind the wheel. Certainly it is far more suited to warm weather, but it is, with the optional side curtains and quarter panels ($45.60) entirely practical for winter motoring as long as you could fit a heater. The longer wheelbase gives it a much better ride than the normal Mini, the seats are very comfortable and resemble those fitted to the Citroen 2CV and the initial Renault R4s.

With an all synchromesh gearbox and high gearing (16.0 mph per 1000 rpm in top gear) it is also able to cruise at 60 mph. Top speed varies enormously from about 65 mph to 80 mph, depending on wind direction and how steep the hill is. Acceleration with the 50 bhp 1100 cc engine is very good for much larger cars can be beaten away from the lights and because of its size and visibility with the hood down, there can be few cars quicker in city traffic.

My only criticism of the Moke is the turning circle. At 36 feet it seems unduly large for a vehicle of this size. The larger turning diameter is caused by the fitting of 13-inch wheels (for greatly increased ground clearance and therefore making it more versatile as a farmer's cross-country vehicle) in place of the original 10-inch wheels. It is extremely rugged and so simple it is hard to imagine anything major ever going wrong with the body, if you can call it that, or for that matter any other part of the car.

It is amazingly practical for the interior can be hosed out in seconds; the bonnet lifts out of its rubber hinges and the side body areas are for the petrol tank and battery. As a sports car it is hard to beat. Because it is fitted with Weathermaster tread tyres the limits of adhesion aren't great—try one with radials though and it is almost as fast as a stock Cooper S around a tight circuit, although the front wheels tend to lock up under very heavy braking, but it is marvellous fun to drive hard.

As a first car for young motorists it is perfect. It should be cheap to run, 35 mpg would be normal and 40 mpg easy to achieve, and cheap to buy initially at $1465 including sales tax (are you really a primary producer?) of $135.30. Australia is now the only country in the world producing the Moke and production this year will be trebled to meet the increased export demand.

ABOVE LEFT: For inclement weather and winter motoring, hood, optional side-curtains and quarter panels are available.

ABOVE: In warmer weather with the top down the Moke is a real fun vehicle.

EXTREME LEFT: Front seats are comfortable, but bring your own padding if you are sitting in the rear.

BELOW: It handles like a sports car, but the turning circle of 36ft. is disappointing.

If you like fresh air, you'll like the Moke. The hood is supported by a tubular frame which can be folded flat by releasing a few clips. It is an extremely simple and quick operation. This is an Australian Moke

They don't come much simpler than this! The central pod houses the speedo and bare minimum of switches. This Moke is a very early British example with a steering-wheel-boss horn push and single wiper

They still make them down under, RUNNING

USED CAR ANALYSIS

"IT'S THE NEAREST YOU can get to a road-going go-kart," the man had said as we buzzed around Battersea with the wind battering our ears and the hood flapping with a staccato rattle. **Hot Car**'s poor photographer, Steve Saunders, was hanging on for dear life, his knuckles showing ivory white against the bright red of the grab handle, as I whirled the little beastie round corners, and all the while Guy Coldwell was sitting smiling in the back and enjoying it all.

The beastie in question was an Australian-built Mini Moke belonging to a company known as **Runamoke** who specialise in selling Mokes and Moke bits. Guy Coldwell works for this company, and I am grateful to him and Ron Smith, the company's founder, for giving up a lot of their time to help in the preparation of this article.

The original Mini Moke was developed with military use in mind. However, it was found that it did not really provide any real advantages over an ordinary Mini, so its military career was not pursued. BMC planners thought that it might have other civil applications, though, such as estate or light agricultural use, and initiated its production. This occurred towards the end of 1964.

The original Moke was based entirely on Mini mechanicals. The engine was the standard 848 cc transverse four, driving the front wheels, and the suspension was of the independent dry type. It had rack and pinion steering. The body was of pressed steel, unitary construction, and from its appearance it was obvious that the designers had cast more than a fleeting glance at the famous Jeep. It was supplied complete with a folding fabric top and side screens.

The interior was as basic as they come—four bucket seats with thin upholstery mounted to a flat platform. The sides of the platform were raised slightly and of box section to provide useful storage space and a place for the fuel tank. A single central pod below the screen housed the speedometer with its integral fuel gauge and idiot lights. This was the Mk 1 model which can be distinguished from the Mk 2 by its single wiper arm and blade and its horn push in the centre of the steering wheel.

Between 1964 and 1968, when British production of the Moke stopped, only 1467 Mokes were sold on the British market and 13,051 were exported. In 1968, production was transferred to Australia.

AUSSIE MOKES

The first Australian vehicles had the 998 cc engine and 10-in. wheels, the latter being changed shortly after to 13 in. diameter. Later, a Mk 2 version was introduced with the 1098 cc engine. Production of the Mk 2 Australian Moke still continues alongside the Mini, despite the recent closure of BL's main factory over there. Production in that country has led to several improvements in the breed—many to cope with some of the harsher driving conditions encountered in the "outback".

Visually, the main difference is the use of larger diameter wheels which now look in proportion to the body. Other less obvious differences include slightly higher ground clearance, greater length, a greater height giving more headroom, better cooling system, rubber protectors to guard the CV joints and an integral sump guard. Inside (or should I say on top?) there is a parcel shelf built in on either side of the speedo pod and a steering column lock. Large Perspex quarter-lights are fitted on either side of the screen, and a towing bracket is fitted as standard to the rear bumper.

The Australian factory also produced a limited number of "Californian" Mokes for the American market, but legislation prevented their export and consequently these found their way on to the Australian market and some to this country through Runamoke. They were fitted with the 1275 cc engine and are much sought after by Moke fans.

Although we are really looking at Mokes from a used car point of view, it is worth noting that Runamoke can, and do, import new vehicles for sale in this country. The going price at the

This is one vehicle where leg room for rear passengers presents no problem. The seats are rather hard. Note the side grab handle

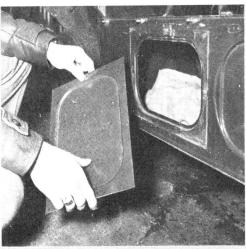
Removable panels in each side of the Moke conceal storage compartments. Not exactly burglar-proof, but then neither is the car

This British Moke uses all Mini mechanicals with the exception of the handbrake cable. Small wheels give a low ground clearance

So have you ever thought of....
A MOKE!

time of writing for an Australian Moke with all import duties, VAT and car tax paid ready for the road works out at £1550.

CHECKS

Now then, what do you look for when you go to buy one? Firstly, on the mechanical side, check the condition of the CV joints and drive shafts. Have a close look at the condition of the hood and side screens as they tend to wear out and a new set is likely to cost in the region of £60. Each set should last four to five years. Finally, rust. Like all steel-bodied cars, the Moke suffers from attacks of the dreaded rust and is particularly prone to the disease in the floor of the battery box, at the base of the rear panel where it meets the floor, along the sills and the front floor panels. The last named can rot through from both sides as, being basically an open vehicle, the Moke can collect water inside, particularly if you've got a badly fitting or worn hood.

The spares situation seems to be quite good, Runamoke carrying a large quantity in stock. British Mokes are Minis mechanically so there is no problem there and most of the Australian body panels will fit the British bodies, there being hardly any British panels left anywhere. The only real difference seems to be in the area of the rear panel, but I understand that this can be made to fit with a bit of ingenuity. British mechanical parts are compatible with the Australian ones, with one or two exceptions, and the oil filter is different. However, these parts can be obtained from Runamoke and if they don't have them in stock delivery from Australia normally takes two to three weeks. Runamoke make their own hoods and side screens which give better visibility through larger windows.

Tuning, if your nerves will stand it, should present no problem because there is a tremendous range of equipment on the market.

PERFORMANCE

Performance is similar to the Mini, what is gained from the lighter weight of the body is lost through the resistance caused by its flat face. Guy tells me that a standard 848 Moke will return an average mpg figure of 40, an Australian 1098 about 45 mpg and the 1275 about 50 mpg. Their top speeds are 70, 80 and 90 respectively; 0–60 times for the 1098 and 1275 are approximately 18 and 14 seconds respectively. With the exception of the 1275, they all run on two-star fuel, which can't be bad in this day and age.

IMPRESSIONS

Guy says the little beasts are quite stable at speed, and the impression gained from our quick blast around the streets of Battersea would suggest this. As far as driving goes they are the same as a Mini, but the body design gives them a lower centre of gravity so that they tend to roll less and are more stable. The ride seems to be slightly stiffer, but this may be due to using standard components with a lighter body shell, and the steering is fairly light and direct. The positioning of the controls could be improved considerably. Steering wheel and pedals are okay, but the central control pod with its switches is out of reach and the gearlever is too far forward and too short.

With the top up, the material tends to flap a lot, and at times the supporting canvas webs that run the length

Mokes are prone to rusting along the sills, as can be seen here. Also check along bottom of rear panel

of the roof can produce a machine-gun-like rattle as they flap. With the side screens off, you get blown about a bit, but not nearly as much as I expected and it is not unbearable.

Basically, the Moke can be regarded as a fun vehicle and I can see how it could be entertaining to drive during the summer. It has certain unique advantages. For instance, with the top off you can practically walk into it, and if you should get stuck in the mud somewhere, it requires little effort to lift the back end and pull it out. The other thing, of course, is that it is so plain that there is tremendous scope for customising.

There aren't too many of them on the market right now and they are sought after, with good British examples going for between £300 and £600. The Australian models fetch more at between £850 and £1300. I would think the Australian model is the better buy, but there ain't too many around and you'll really have to hunt if you want one. Best bet is to try Runamoke, whose address is 73–77 Battersea Church Road, London, SW11 and see what they have in stock.

I.P.

COMING SOON

May's Used Car Analysis will cover all TVR variants. If you have ever owned a used TVR, we want to hear about it! Send your letter soon, marked "Used TVRs," telling us:
(1) The year and model of your TVR ....
(2) When bought, when sold, what price ....
(3) Nature and cost of any repairs .....
(4) Average fuel consumption ....
(5) Whether it was modified ....
(6) Would you buy another? ....

SKD Moke

(Semi Knocked Down)

Suggested sub-assemblies for easy construction with a minimum of plant. Any variations can be accommodated.

BACK TO BACK
MAKING TRACKS

Ironically, Citroën introduced the go-anywhere, do-anything Méhari just as BMC withdrew the Mini Moke from the British market. Peter Nunn compares these two open-top cult machines

In the beginning they were both designed as basic, go-anywhere utility vehicles, each being the kind of machine, say, Farmer John (or Jean) would use down on the farm, to carry essential supplies, or even livestock come to that. But BMC's Mini Moke, in company with Citroën's Méhari, soon took on another, completely different role; it became fashionable – both to drive and to be seen in. Today, the Moke thrives in this Kings Road context, having long ago abandoned any commercial vehicle pretensions it may once have had. And the Méhari? That, too, flourishes – but not in Britain, strangely enough as it's never been officially imported by Citroën Cars Ltd, in any shape or form.

Austin-Rover, or BMC as they were then, summed up the Moke's capabilities quite simply in an early factory hand-out. Described as a 'rugged run-about with a thousand uses', the Moke, the British company reckoned, could serve as a hotel beach wagon, holiday camp taxi, golf course caddy truck and building site survey vehicle; farmers, estate managers and vets would also find a use for the vehicle, the catalogue reasoned. In promoting their utilitarian Méhari, Citroën adopted a similar approach, picturing it loaded down with straw bales, market products (baskets of fish, boxes of flowers and so on) and hard at work carrying barrels of oil at the refinery. It's significant, though, that the latest brochures for the two cars (yes, the Moke is still being made, 15 years after it was dropped from the UK market) show the duo skidding and playing about on the beach, as open-top fun cars for the young-at-heart. It's this enviable combination of down-to-earth practicality and dune-buggy-style character that typifies both Méhari and Moke.

Labelled by Citroën as 'sobre, solide, sure, devouée, capable de tout', the Méhari acquired its name from the Sahara dromedary adopted by the Touareg Arab tribesmen. In essence, this light-weight Citroën pick-up was Dyane based;

The 'roo bar' was an option on Australian-built Mokes

Typically Mini – the Moke's 'cabin' is simple but comfortable

A familiar sight – the A-series power plant of the Moke

that is, it was constructed around a toughened version of the familiar 2CV/Dyane platform chassis whose origins stretched back to the early thirties. Independently-sprung back and front by means of inter-connected leading and trailing arms, this chassis, in Méhari guise, supported a tubular framework which, in turn, located the vehicle's unusual body panels.

Citroën decided to use a hitherto little-known manufacturing process (little-known in wide-scale, mass-produced automotive terms, to be more precise) in designing the Méhari's corrugated bodywork. The panels, 11 of them in all, were made using ABS, the initials standing for Acrylonitrile Butadiene Styrene. Briefly, these thermo-plastic body pressings are self-coloured, easily repairable and extremely tough, the manufacturing process (based on petroleum and coal by-products) also making them easy to keep clean and, by all accounts, scratch-proof.

In styling the Méhari, Citroën came up with something of a masterpiece. For, not only does it have the option of cut-down doors and a fold flat front 'screen, it also has twin bonnet straps to keep pre-war sports car drivers happy. On a more serious level, there's a choice of two basic versions – a two-seater with flat rear deck section, or a 2+2, utilising an additional pair of rear seats.

Citroën's inimitable 'hair drier' flat-twin, displacing 602cc, powers the Méhari, the latest versions producing a modest 32bhp (DIN) at 5750rpm. The first Méhari, introduced in May 1968, turned out a meagre 25.5bhp in comparison. This all-alloy, air-cooled ohv engine, driving the front wheels (naturally), is coupled to a modified four-speed manual 'box. Brakes are inboard at the front – discs replaced the large drums in '78 – and outboard at the rear; the handbrake works effectively on the front wheels. Steering is by rack and pinion while the suspension consists (as already mentioned) of front leading arms, trailing arms behind, twin chassis-mounted helical springs and telescopic dampers.

The Méhari has spawned several interesting off-shoots over the years, the most notable perhaps being the four-wheel-drive variant, known simply as the Méhari 4×4. Identifiable by means of its distinctive headlamp grilles and bonnet-mounted spare tyre, the 4×4 boasts no fewer than three facia transmission levers and up to seven forward gears.

Currently a 4×4 costs the equivalent of £5502 in France (barring taxes) whereas the less complex two-wheel drive Méhari is priced at £3231.

In case you were wondering, incidentally, why Citroën have never imported the Méhari into Britain, the reasons are threefold. First, there are complications in the job of converting to right-hand-drive; second, it would be difficult, if not impossible, to have the Méhari Type Approved; and thirdly, (and most importantly) the cost of conversion, Citroën say, would not be economically viable in view of the limited market in the UK for vehicles like the Méhari.

If you need proof that the UK market is regarded as limited, you need look no further than the success of the Mini Moke – or rather its lack of success. The Moke started off as a prototype farmer's hack, a sort of Mini Land Rover if you will. The first Moke had a simple spartan tub, two seats and a windscreen. Up front was the by-now ubiquitous 850cc Mini engine, transversely mounted, and driving the front wheels.

Ignoring the body, however, the biggest difference between the prototype Moke and the everyday Mini, was that the Moke had another engine in the rear driving the back wheels. The Land Rover had four-wheel-drive, so the Twini-Moke as it was known, should have it, too.

Alas, it didn't work. Despite many development miles the twin-engined Moke was destined to be a blind-alley development. The fact that John Cooper had a horrendous accident in a twin-engined Mini at around this time further underlined the problem.

Doomed from the outset

So when the production Moke finally appeared – in mid-1964 – it had but one engine and but two-wheel-drive. It was further hampered by a lack of ground clearance and that, despite claims from BMC, it was never classed as a commercial and thus failed to attract tax concessions. It was doomed from the outset. Of a British production run of just less than 30,000, fewer than 1500 were registered in this country.

It was in 1968 that production transferred to Australia where it continued until only recently. In Australia the Moke was greeted as a long-lost friend and, for some reason, the lack of four-wheel-drive did not seem to matter too much. Changes were made to the little car. The puny 10ins Mini wheels were thrown away to be replaced by 13ins examples, helping ground clearance in the process, while the 'car also grew a little. To accommodate the slightly fatter tyres, an inch or so was added to the wheel arches front and rear, along with plastic mud flaps, and stout perspex side screens were added to the 'screen assembly.

Gradually over the years further improvements were made. The hood was given zip-in 'doors', the

The spare on the bonnet means this is a 4×4 Méhari

The twin-engined Moke prototype made for the US army

848cc engine gave way to the 998cc A-series unit, and the 1275cc engine was an option. And the car became a cult machine. Roll-over bars were added, even wider wheels, nudge bars front and rear and so on.

In Britain, meanwhile, enthusiast Ron Smith felt there was now a market for the Moke so he started to import them. Since 1980 he believes he has sold around 200 . . . though supply from Australia has now all but dried up. See the news pages for details of Ron's, and Runamoke's, latest plans.

So tracking down a Moke for this feature was easy. A phone call to Runamoke in Battersea and a 1275cc 'Californian' demonstrator was instantly, and obligingly, made available to us for a day. In the interest of originality and all-round fair play, it might perhaps have been wiser for us to have tried one of the 998cc 'standard' Mokes as the poor old Méhari has to make do with just 602cc and just 42bhp. To this end our Specification Table covers the less potent, yet more common-place 1-litre Moke. But we're grateful nonetheless to Ron Smith and his team for providing the Californian at such short notice.

No, finding a Méhari was the real problem. We tried virtually everyone and anyone we knew, but to no immediate avail. Eventually, purely by chance, came across Noel Balbirnie's orange Méhari parked in a London side-street. After a little gentle persuasion (!), Noel bravely agreed to take part in this story. Our thanks, therefore, go out to him as well.

It was while Noel was in Holland three years ago that a newspaper ad for a fairly early Méhari two seater appeared. To cut a long story short, he ended up buying the Citroën for just £370 which, considering the majority of Dutch-domiciled Méharis fetch between £800 and £1000, seems quite a bargain. The car is, nevertheless, quite sound – its plastic body (it was originally green) having stood up to the rigours of the road well.

'Officially', his Méhari dates from 1973 but it could, in fact, have been built prior to that as its rear suspension carries the inertia dampers fitted to pre seventies 2CVs and Dyanes. Its speedo is currently reading 56,000km. Since 1979, Noel has replaced the gearbox to eliminate a fourth gear whine and substituted an Ami twin choke carburettor for the original. And because the Méhari comes without seatbelts (a remarkable state of affairs in this day and age), Noel has devised his own lap-and-diagonal set-up which looks rather suspect but at a quick glance seems strong enough.

The Méhari's tent-like hood peels off in sections to reveal, well, not very much at all. When stripped of its voluminous rear covering, there doesn't really appear to be a great deal to the car; sure, the interior is very spacious, hard-wearing and practical but it's also spartan and scantily-equipped, with few cockpit dials, gauges or controls, for example.

'Our' Californian Moke, meanwhile, came with all the trimmings. Externally, one can't fail to notice the hefty cow-catcher nudge bar at the front, nor the susbstantial roll-cage that's almost as high as the car long. The high-backed, twin seats also figure prominently as do the englarged 13ins wheels shod with 175 section Dunlop Weathermaster tyres.

In comparison with the Méhari, the Moke's dash seems almost over-endowed with switches and instruments galore although in reality it's no better equipped than the basic Mini City saloon. On the floor, the Moke has swimming-pool-style matting

Méhari has twin bonnet straps and a fold-flat screen...

Spartan with no frills, the interior has zero legroom

Citroën's inimitable flat 'twin' produces just 32bhp

and at the back Runamoke's own design of lockable cabinet; but the latest type of inertia seat belts (securely mounted on the Moke's stout roll-cage) do provide a most reassuring sense of security – Méhari drivers take note...

While the Citroën's awning may take five to ten minutes to dismantle completely, the Moke's hood comes down simply and effectively in virtually no time at all. And it folds away neatly in one go, too, something that can't be said of the Citroën canvas. Both cars have full sidescreens to keep rain and buffetting wind at bay. Doors, though, are optional on the Méhari; incredible though it may seem, it's possible to buy one new with just simple chains across the door apertures. Only the French could think of that one.

To drive, the cars are, as expected, poles apart. The Moke, in 1275cc form, positively flies off the line, cornering quickly in the bargain thanks to the famed Mini roadholding capabilities and, of course, those fat SP44 tyres. Well-balanced and predictable, the Moke 1275 is acceptably fast on the straight and super-controllable through bends, understeering strongly yet safely at all times with little roll. Ride comfort is good, the Moke being, overall, an easy and untiring machine to drive – whether it be through Chelsea on a sunny Saturday afternoon or on a hard, press-on trip across country.

If anything lets the Moke down while on the move it has to be the transmission. In familiar Mini fashion, the gearchange is habitually ill-defined and sloppy, the drive-shaft UJs chatter and there's still that annoying transfer gear rattle. And the Californian's gear lever is a little too short, by about 4ins.

The Moke's steering, furthermore, despite its welcome 'feel' and two-and-a-bit lock-to-lock directness, has a pretty uninspiring lock. Braking is certainly up to the job in hand, although nothing special. Entry and exit, effected by gingerly negotiating the pannier box members on either side, also needs some getting used to.

The Méhari is something else altogether – literally! For a start, as there's no legroom to speak of, six-footers and over can forget straightaway about ever driving one comfortably. If you've even driven an Italian sports car that has its steering wheel far too close to the pedals, you'll appreciate the problem. In other words, get ready to *benz ze nees*...

Aside from a lowish driving position and large, bus-driver steering wheel (both of which tend to exacerbate the above situation), the Méhari behaves pretty much like any normal 2CV or Dyane on the road. Of course, depending on your definition of the word 'normal', that could be a good or bad thing.

Slow though it is, the Méhari isn't that far adrift from the times put up by the 998cc Moke tested by one Australian motoring magazine, back in the late sixties. On top speed and acceleration through the gears, the Moke is clearly superior. Yet the Méhari can still be coaxed along at quite respectable speeds, 65mph at 40mpg being possible all day and night with only the minimum of fuss and bother. True, 0-60mph is lethargic but neck-snapping performance figures are not what the car is about.

Legendary 2CV eccentricities

Knowledgeable Deux Chevaux people will tell you the Méhari is a difficult car to drive. Noel Balbirnie reckons his example isn't as *nice* to drive as a 2CV but that's about as far as it goes. Certainly, BRK 630Y has all the legendary 2CV eccentricities. It rolls tremendously around corners, it has the wonderful push-pull gearchange that so many people find off-putting until they actually try it – and, from the outside, it looks 'funny'.

But to temper all the body lurching, understeer, groans, squeaks and lack of torque – the engine is noisy, too, say the critics – comes a superb ride and an unexpectedly high level of refinement. Once mastered, the Méhari should be an extremely civilised form of transport – and it's fun to drive as well.

Summing up the Méhari's good points one can say it's incredibly practical, cheap to buy and to run and, an important advantage this, in today's rationalised world – it represents 'something different'. Plenty have been built by Citroën (some 132,000 to date) so finding one to buy abroad ought to be no problem. Not only is it refined and capable far beyond its station in life, it is, without doubt, much more of an 'off-road' vehicle than the Moke.

On the debit side, the terrible driving position, the bouncy road manners (both of which need acclimatisation) should be mentioned. Pre-1978 model-year cars will have the controversial front drum brakes but seat belts could be a headache on all versions. Goods carried in the rear are all-too visible and vulnerable to theft while spares (especially for the unusual body) might be difficult to obtain in a hurry. And motorways with nasty head winds, not to mention steep hills in general, can make Méhari driving a misery.

As for the Moke, it benefits from known Mini mechanicals – hence replacement parts and service will be cheap and easy to locate anywhere in Britain. It's a tougher, more conventional (sic) kind of car than the Méhari which, in turn, is more economical with fuel, but little match on the performance front. Fast, energetic driving similarly favours the charismatic Moke, one of *the* sixties cult machines, immortalised by *The Prisoner*, Carnaby Street and others.

Conversely, while the Moke suffers from all the Mini's foibles – drive-shafts, gearboxes, rear sub-frames to name but three, getting in and out can also be trying at times. A Moke Register has recently been started (see 'Club Focus'), but the Méhari comes under the auspices of the nutcase 2CVGB enthusiast group and the all-enveloping Citroën Car Club, a situation that must make Méhari ownership more practical and enjoyable.

As far as appearances are concerned, office opinion had it that the original Mokes, always very thin on the ground in Britain, were better looking than the pricey and rather tarty Californian breed currently available. That said, Runamoke's Californian was a bundle of fun, creating a great deal of attention wherever it went.

Conclusions? If you appreciate the finer points of Citroën 2CVs, you'll plump for the Méhari as a matter of course. Should Deux Chevaux hold no fascination, the Moke must get the vote as a 2CV is one of the archetypal love/hate cars. On a personal note, I'd settle for a Méhari with decent legroom – but then I owned a yellow 2CV6 for a year and half not so long ago, and loved it...

SPECIFICATION	Mini Moke	Citroën Méhari
Engine	In-line 'four', water-cooled	Flat 'twin', air-cooled
Bore × stroke	64.58mm × 70.61mm	74mm × 70mm
Capacity	998cc	602cc
Valves	Pushrod ohv	Pushrod ohv
Compression	8.1:1	9:1
Power	39.4bhp (DIN) at 5200rpm	32bhp (DIN) at 5750rpm
Torque	50.4lb.ft (DIN) at 2500rpm	30.5lb.ft (DIN) at 4000rpm
Transmission	Four-speed manual. Front-wheel-drive	Four-speed manual. Front-wheel-drive
Final drive	4.2:1 ratio	3.87:1 ratio
Brakes	Drums front and rear	Drums front and rear
Suspension F.	Ind. by unequal length transverse arms, tie-rods, rubber springs, telescopic dampers	Ind. by leading arms, inter-connected coil springs, friction dampers
Suspension R.	Ind. by trailing arms, rubber springs, telescopic dampers	Ind. by trailing arms, inter-connected coil springs, friction and inertia dampers
Steering	Rack and pinion	Rack and pinion
Body	Monocoque, all-steel	Steel chassis, ABS body

DIMENSIONS	Mini Moke	Citroën Méhari
Length	10ft 0in	11ft 6in
Width	4ft 3½in	5ft 2in
Height	4ft 8in	5ft 4in
Wheelbase	6ft 8in	7ft 9⅓in
Kerb weight	12¾cwt	10⅓cwt
Tyres	5.60 – 13	135 – 380

PERFORMANCE		
Max speed	84mph	65mph
0-60mph	27.9sec	30sec
Standing ¼ mile	23.5sec	22sec
Overall fuel con.	40mpg	45mpg
Years built	1964 – date	1968 – date
Nos built	29,393 (1964/68)	132,000 to date
Price now (exc. taxes)	£3627/£3797	£3231/£5502

Note: Moke specifications relate to 'standard' model. 1275cc, 54bhp versions have front discs and 175 R 13 tyres. Californian spec 998cc Mokes weigh in at 13¾cwt, 1275cc versions at 13⅞cwt.

STORY — RAY RYAN. PICS — DES KINSELLA.

We poke the Moke

"Moking may not be a wealth hazard but it can be very boring".

Spending two weeks with a test vehicle is a vastly different proposition to actually owning the thing. That vehicle has not cost the tester one cent; whereas any member of the Great Australian Motoring Public will be paying top dollar for that same privilege.

Pre-purchase test drives are usually little more than a fully-escorted squirt around the local backstreets in the smiling company of Bill Crunchouse or one of the other 'stitch-em-up boys' from your local dealership. After that you sign some papers, pay out a fistfull of money and whoopee . . . you've just bought yourself a motor car. It often takes many months for a driver to establish if that vehicle was a good investment and whether it fulfills its advertised value-per-dollar claims.

We regard the latest version of Leyland's 'Moke' as a classic test case of the vehicle that just isn't worth the money, as despite the over protection offered by the Federal Government (in the form of tariffs and restrictions) the locally-built Moke is simply way too dear. No doubt some Leyland marketing exec will attack us with a, "The Moke is still the cheapest vehicle on the local market . . . etc!", and our reply is, that if that is the case then the vehicle is still far too costly in terms of actual value for money.

Simply being "cheaper than the others" does not essentially make the Moke either attractive or the Bargain of the Year.

A basic Moke (less side curtains) sells for some $3,100 with the dressed-in-denim-vinyl version, or 'Californian' which we tested, going out the doors for around $3,600 prior to actual on-floor horse trading.

In late 1972 the writer purchased a new 1275cc Moke Californian complete with radial tyres and side curtains for around $2,100 from a major Sydney dealer. In that last five year period the Moke has leapt in cost by $1,500 (on the Californian), an increase of some 70% and now offers far less in terms of either performance or inherent value.

Even when the impact of inflation and the cost burdens of ADR compliances are taken into account, the Moke still shapes up poorly. Other popular makes have all shown some degree of engineering and design development which can partially offset the impact of the rising costs of Australian-built machinery, but the Moke has gone the opposite way. The current 998cc engined Moke is a great leap backwards when compared to earlier 1100 and 1275 models and the development/improvement has been overlooked in even the most basic of areas.

The added $500 worth of 'Californian' pack doesn't really alter the value-per-dollar situation. It includes white spoker wheels, denim hood and seat trim (in vinyl), a neat little leather rimmed wheel, bar work at both ends and a few decals. That's basically the story and the package has little real value apart from the wheels and does little to really improve the vehicle despite the added cost. In our opinion the basic Moke should begin with the Californian and then be improved from that basis. The retail price is way too high and the improved Californian version should sell for no more than $3,000, including side curtains.

Given a free hand it would not be difficult to develop the Moke into an ideal recreation and fun vehicle, but Leyland appear to have missed the boat completely and the current offering is still plagued by the inherently weak areas which have been a problem with the Moke since its inception.

Lack of predelivery preparation has been a constant criticism levelled at the Leyland product and unfortunately the 'Bushdriver' test machine was no exception. Our Moke (licensed in NSW) was delivered to our west coast test team with gleaming paintwork and virtually negligible mileage on the big central speedo. Even the fuel tank was topped up to the F stop and things

looked promising initially. However, clutch adjustment was so far off specification that we had no more than ¼ to ½ in of travel before the clutch engaged completely. It had to be depressed completely to the floor before it would free entirely and on more than one occasion we were caught in the city traffic with a vehicle that wouldn't slip into any gear!

A total of three wheel nuts were loose and one of these was actually only finger tight! The spare wheel lock and security nut was loose and finally fell apart while both the horn and handbrake warning light were completely inoperative. Neither worked at any stage during our testing of this brand-new vehicle. Both outside rear view mirrors were poorly fitted and had to be re-tightened to prevent them from shifting out of adjustment at anything above 40 km/h. Brake fluid slopped from a cross-threaded master cylinder cap and ran onto the firewall paintwork, while both brake and clutch pedal rubbers fell off after only three kilometres of use.

How would you feel if you'd just paid $3,600 for your Moke and it performed like that . . . before you even had the chance to get it dirty?

Originally the Moke was available with both the 1100 and 1275cc Cooper-derived motors; reasonably revvy transverse fours that would shift the mighty fun cart along with the best of the sub-two litre family machines.

That was way back then. Right now, the Moke and all its Mini-based brethern are only offered with one motor; a 998cc with all the mandatory corks, plugs and power sappers as dictated by ADR27a. Even in pre-ADR times the one-litre BMC motor was regarded as somewhat of a bad joke and with the added hassles of detox plumbing the motor can now only be treated as a dismal failure and an antique that is best forgotten! This pathetically weak motor hampers, spoils and destroys a vehicle which has sufficient potential to be one of the greatest all-round fun things of the decade.

Working on a Moke engine is still an enormous bitch. Drop a spanner down inside that engine bay and it's gone forever, with the chances being that it'll never even hit the ground! General access to the engine bay (NOT the engine components themselves), is however still first class as the Moke's bonnet still slips completely free from the vehicle, being attached with only two metal/rubber clips at the rear bonnet edge. Two moulded rubber 'rally clips' give added bonnet security up front and attach to the forward sides; a cheap but effective setup.

The distributor has received a slightly larger version of the original "useless plastic cover" and the new effort is as totally ineffective as ever. It will keep away the usual rainy day rainwater that you'll encounter on the way to the office, but as an effective water shielding device for even the most casual bit of bushdriving it's completely hopeless. The Moke motor bursts into life with a little choke when cold and will fire on the first crank of the key when hot. It runs cleanly without the usual pinging associated with many detoxed motors and runs a constantly amazing consumption of over 40 m.p.g. It would run forever on the faintest sniff of an oily rag but this consumption figure is still NOT BETTER than that achieved with a normal, pre-ADR, 1275cc model. The new motor is still economical, but gutless! It runs cleanly under load but still persists in running on after it's been switched off and in many ways it is the ultimate motor of paradoxes. Having looked back on earlier Leyland vehicles and considering such efforts as the rally Coopers, Evan Green's trans-Sahara P76 and even Tholstrup's Moke, we know that Leyland CAN do it when they really want to. This year they didn't even try!

The gutlessness of the Moke's one-litre mill is not only a hassle in the bush,

but also on the road where you'll find yourself locked into a life and death struggle with thundering Datsun 120Ys, Mazda 323s and tricky Gemini Sandpipers . . . the real performance iron! Pull out of your driveway in the morning and bury your foot into the Moke's firewall as you cross yourself and mumble a few chants. There's so little acceleration on tap that you can only time your progress with a calendar as the Moke moves slowly up to around 60 or so, and the traffic locks up and banks up behind you.

In the dirt it's just as bad and the poorly spaced, wide ratios of the Moke's four speeder only make things worse. Closer ratios are required to take any advantage of this motor's attempts at producing any power, but the Moke is still fitted with the stock, wide-ratio item from the Mini saloon. Stoke the motor up and snick into first and you're rolling through the bush in your Moke. Snatch second and the power dies; a combination of zero muscle and a set of ratios that are only suited to a larger or far more torquey motor. This little machine screams for a closer gearbox as well as a real motor that will actually do the job.

The now-extinct 1275 Leyland motor was no world beater in stock trim, but even with a single SU carb the earlier Moke was still capable of either towing a light trailer or carrying two or three humans and a fair bundle of gear along at a reasonable rate of knots. The one-litre motor began back in the past with the Morris Minor, and in our opinion that's where it should have stayed. To reintroduce it in 1978 makes as much sense as collecting EK Holdens or installing a Land Rover six in your Overlander!

On smooth or mildly rippled dirt tracks and forest trails the Moke still is one of the most pleasant riding vehicles that you'll ever try. With a good motor and gearbox combination that pleasure would increase proportionately, but nevertheless we found that everyone who drove the Moke in the dirt came away as a convert to the front wheel drive concept. The 'Californian' doesn't pitch or sway and the time-tested and rugged rubber cone suspension gives the type of comfort that only a fully independent setup can deliver. Despite shocks that are purely marginal the Moke never gave our test drivers any hassles but we all found ourselves asking: "Hey, if it's good now, then just imagine what it would be like with some really good shocks on board?"

The Moke handles as all Minis have ever done, with a throttle-conscious understeer that's easily controlled. The main limitation to correction is the feeble nature of the motor which lacks the punch to really straighten things up when some muscle is called for. Just

remind yourself that it's no 1275 before you set loose to play Rally King on your nearest square of available earth. Steel belted 'Rally Radials' are standard wear on the five inch ROH spokers fitted to the Californian, while the cheaper Dinky Toy version receives only a set of skinnies and the usual crossply rim protectors. Some dealers will try and burn you an extra $100 or so for the added merits of a set of winter treads which are almost useless for anything but a trip across the Scottish moors. The ROH wheel and tyre combo should be the stock equipment on any Moke, regardless of price tag.

Handling is greatly improved both on the road and in the dirt, with the new rim and rubber combination giving any driver both greater confidence and more positive feedback through the highly geared steering and tiny leather wheel. In loose sand or deep mud however, the Moke is still no better than ever and is a constant embarassment to any driver who's out trying to keep up with his friends in their 4WDs. Earlier Mokes with more power should claw themselves through the softer stuff with a constant throttle opening and a fair amount of wheelspin, but the lack of muscle of the current one-litre has ruled out that line of work. With the little motor it becomes a red line torture exercise all the way and even that usually ends up with the Moke bogging itself into the soft ground like a baby mole. If you keep the revs up and the tyres pressures down, it's possible to plane the Moke over a soft surface but all that's needed is a slight hesitation or one stop, to read the terrain, and the Moke will bog itself under when you try to get rolling again.

Loose gravel tracks, moderate hillclimbs and idyllic morning drives in cigarette commercial country are where it's at with this year's Moke. In the dust and dirt the Moke is a lightweight, well-suspended fun machine that will deliver the goods as long as you don't expect too much in the way of serious ORV potential. Despite the looks and styling, it's still NO 4x4 and a person would be crazy to consider one as an off-road alternative to either the F20 or LJ80. However, as an option to any second-hand Datsun, Mazda, Toyota or similar family vehicle, the Moke offers a slightly broader scope of activity than that of any conventional family four.

If getting going was a problem with the Moke, then the braking system must constitute nearly another 50 percent of all that's inherently wrong with the machine. Leyland's boffins have shaped the metal pedal pads into a cute, semi-diamond shape and then protected them with those rubbers that fall off so easily. With the location of foot pedals, steering column and footwell sidewalls there's just NO way that an average size eight will squeeze neatly onto that brake pedal. If you're wearing thin sneakers it's just possible to hit the thing, if you tilt your foot at some 90 degree angle and occasionally look down to see where your feet are headed! When you're wearing boots or conventional shoes, the problem is insane. The welt of the boot catches on the protruding metal edge of the brake pedal and traps your foot . . . lights flash, brakes go on and off and engine revs rise and fall as you whistle down the road to a half stop. Good fun. It's a disaster and we're still trying to figure out exactly how the Leyland team managed to score their compliance plates for the Moke with this type of design inherent.

Even if you've managed to hit the pedal there's no guarantee that the Moke will stop in a hurry. The new Moke has near useless brakes. It is impossible to even approach the deceleration power of many ten year old vehicles and the Moke has neither the braking force or pedal power to ever manage a full, 'lock-up' stop. You sort of mash the pedal to the floor and then sit back counting sheep as the Moke drifts slowly to a halt. The non-power-boosted drums have not been improved in any way over the years of production of the Moke models and in fact they now shape up even worse when the volume and pace of traffic flow is considered. Minis can be made to stop, so why not the Moke? Perhaps no-one ever bothered to try and cure this problem.

It seems that Leyland have really done nothing to improve the Moke in anything but the most cosmetic manner and many problems that we encountered a few years ago are still there, and are still as annoying as ever.

The side screens still snap loose from their "Lift the Dot" fasteners at two points along their upper edge, and still don't secure properly with their wire-in-rubber locks along the bottom. At anything above 30-40 km/h the screens flap loose and the only comfortable way to drive the Moke is with the top up and all four side curtains removed completely. People still call the Moke by it's old name . . . 'The Joke'. It's easy to see why.

All other manufacturers have incorporated some changes and many improvements in their product but the Moke has been completely forgotten by its creators, and the requests (if there ever were any) of Moke buyers have been totally ignored. Surely after all these years there is SOMEONE who can at least relocate a few 'Lift the Dot' fasteners, so that they don't tear loose every time you take off down the road?

While that person was carrying out his improvements he could probably turn his hand to a few other items. He could puff some powdered graphite into the ignition/steering lock and hopefully loosen it up so that anyone could use it. The lock on our test vehicle required Herculean strength to withdraw the key after the thing was switched off. He might also redesign the heater ducts feeding to the windscreen and possibly devise a neater and more durable way of finishing the joins in the hoses. Staples may look fine on the paperwork you leave scattered about a desktop, but on the demister hoses of a $3,600 motor vehicle they don't really make it!

If we didn't seriously like the actual Moke concept then this test would have been considerably easier, as we could have given up and simply filed the machine in the 'H for hopeless' drawer. We would have been admitting that the vehicle had no real potential, and that potential is good, old-fashioned F-U-N; a commodity which the Moke can deliver in abundance. If Leyland's marketing team ever pull their heads out of the sand and took the time to realise this one point, then they might bother to allocate some time and design team talent to working the Moke into a feasible RV package . . . at a realistic price. It can be done and should be done, but we doubt seriously if they'd even bother to take up the challenge.

MISCELLANEOUS MUMBLING:

The Californian is fitted with genuine 'public shower' style rubber matting to protect the paintwork in both front and rear footwells, and also offer somewhere to wipe your boots on a muddy day. The front mat lifts clear easily, allowing the owner to sweep or vacuum the crud away from beneath, but the rear matting is trapped in place by the seat belt retaining bolts. To remove the matting entails that you first completely remove the seat belt attaching hardware; a move that doesn't really make a lot of sense. Large rubber grommets are provided in the drain plugs of the floorpan, but those on our test vehicle were such a tight fit, that they were almost impossible to remove to drain some suds from the floor area. Once removed with a screwdriver blade, they were equally as difficult to refit after they'd been lubed with a smear of Vaseline.

A foldup zip fastened rear window in matching denim vinyl trim is fitted to the Californian. Nylon zips keep the thing in place and work exceptionally well, even when spattered with mud; but the rear-mounted spare wheel assembly must be completely removed from the vehicle to enable the lower flap to be secured to the bodywork. A person with very small pinkies, or a child, MAY be able to clip the window material into place but any driver with average sized hands will find it impossible. Access to this rear vehicle section is good via the rear window, but verges on impossible from any other direction when the side curtains are in place. The side curtains are a hassle to remove quickly and cannot be fastened from within the rear of the Moke. Its always up to the driver to play chauffeur and lock all the passengers into place as he runs around the car, sealing various flaps, tying down curtains, slipping zips and of course, 'Pressing the Dot'.

The Moke wiper motor features the mandatory two speed setup but is completely lacking in power and began to expire after no more than five minutes of intermittent use in a sudden Perth summer storm. It would stir into life and then bog down, indicating that there was something amiss with the motor of our particular car. We felt nevertheless that this should have been detected by the quality control department at the factory. The heater/demister unit packs a reasonable punch, although most of the benefit is lost when the side curtains are down, but it delivered no more than 20 percent of the total air volume to the passenger side. It's usually a 50-50 split to both sides of the screen, unless the Moke was harbouring some secret technology which we'd failed to detect?

The formerly cranked gearlever of older Moke models has been replaced by a plastic-wood topped "sporty straight" item that doesn't twist toward the driver. To swap cogs it's necessary to lean forward in the seat and a lot of feel is lost through the lever itself. It also looks very cheap and fragile and the plastic knob slips easily in your palms. Yuk!

Metal 'Moke' badges are riveted to the rear mudflaps and are usually torn off during the first run in the bush. The rear flaps are drawn up over the wheel when reversing and the nameplates soon lose themselves. The flaps however, on all four wheels are an excellent idea, and the fully enclosing front items (rubber) kept us dry and protected the driver from flying gloop when we were playing about in the slop.

Comments on the fuel tank cap location and design were somewhat mixed. The writer liked the idea and the ease and simplicity of the whole thing, but two other members of the team felt that it was both dangerous and poorly engineered. Both believed that the excess fumes from a refill were a potential fire hazard and that there was always the risk of a service station operator slopping fuel over the driver.

POSITIVE POINTERS:

We loved the handling, loved the all-round visibility, suspension, steering and manouverability of the Moke and could have even fallen in love with it if we weren't doing our job. With the handbrake on and your right foot stuffed into the loud pedal, it'll spin 360s within its own length and will continue to do so all day. This little trick works beautifully on narrow trails that suddenly turn into a dead end, and is a ball when you're out to prove just how stable the basic design really is.

It'll fly off, jumps flat and solid, just like a good dirt bike and will pick a path through fallen logs and around the rocks with the ease of a lightweight 4WD. In town the Moke will also back up by handling the daily drudges of commuting and shopping with reasonable comfort and superb fuel economy. In the city on a hot day there's nothing quite as comfortable to drive as a Moke with the sidecurtains removed. It's the cheapest and most effective form of air-conditioning that we know of.

As people who are into fun things, we loved the Moke. Friends, girlfriends, dogs, lovers and relatives all dug the thing and said that they'd like to own one . . . "if only it wasn't so dear" or "if only it went a little better". Get the message? As test writers and motornoters we were shocked at the total lack of performance, the poor braking and general lack of improvement in this model. The sump is protected by one of the best standard-equipment sumpguards that we've seen and the quality of trim and paintwork is well up to the standards of many more-expensive machines. Driving position, general ergonomics (apart from that brake pedal) and a feeling of driver/vehicle integration are all top class with the Moke. It's the kind of machine that feels like an extension of your own body, and as our former editor (Mr Flynn) once said: "It fits . . . not fights!"

The Moke has enormous potential, but potential isn't what you're looking for when you pay out over $3,500 for a new vehicle. As is we'd recommend the Moke to any reader, but NOT a new, emissionised one-litre Californian. We'd suggest an older 1275-engined version which could be picked up very cheaply on the second hand market and then transformed at a reasonable cost. A few repairs, some cosmetic changes and minor improvements would turn the little machine into an eye-stopper which would outpose, outperform and outvalue any optioned-up variant from the Leyland folks.

The current Moke is a sad case of "We like it personally for some good basic reasons, but we wouldn't like to have to buy one." Things are always different when you haven't had to dig into your pocket, just to find out that you may have been wrong all along!

TECH SPECS
Leyland Moke Californian

DIMENSIONS

Overall length	3,232 mm (127.25 in)
Overall width	1,450 mm (57.00 in)
Overall height	1,600 mm (63.00 in)
Wheelbase	2,095 mm (82.50 in)
Track, front Stock Moke only.	1,447 mm (49.00 in)
rear Plus 0.5 in on Californian.	1,466 mm (49.75 in)
Load deck size	Not applicable.
Ground Clearance	203.2 mm (8.0 in)

WEIGHT

Curb weight	611 kg (1344 lbs)
Weight distribution	Not available
Gross vehicle weight	991 kg (2,180 lbs)
Seating capacity	2 persons
Maximum payload	300 kg (660 lbs)
Rated towing capacity	400 kg (880 lbs)

ENGINE

Type 99H	Four-stroke cycle, water cooled, transverse located. Valve operation via pushrods.
Number of cylinders	4
Lubrication system	Wet sump
Bore	64.58 mm (2.543 in)
Stroke	76.2 mm (3.00 in)
Capacity 998cc	(60.96 cubic inches)
Compression ratio	8.2:1
Carburettor	1 x SU type HS4.
Air cleaner	Paper element type
Maximum power	30 kW (40 BHP) @ 5,100 rpm
Maximum Torque	69 Nm (51 ft/lb) @ 2,600 rpm

ELECTRICAL

Ignition timing	7° BTDC @ 1,000 rpm
Standard spark plug	Champion N9Y
Contact point gap	0.35 to 0.40 mm (.014 in to 0.16 in)
Generator	Alternator
Battery	12V-40 A.H.

TRANSMISSION

Clutch type	Dry, single plate
Transmission type	4 forward, all synchromesh. 1 reverse
Overall gear ratios	1st 15.0:1
	2nd 9.41:1
	3rd 6.09:1
	4th 4.426:1
	Reverse 15.08:1
Road speed per 1,00 rpm in 4th gear	25.7 km/h (16 mph)

POWER CLAIMS

Maximum power	30 Kw (40 BHP) @ 5,100 rpm
Maximum torque	69 Nm (51 ft/lb) @ 2,600 rpm

STEERING

Turning circle	9,754 mm (32.00 feet)
Toe-in	Not quoted
Camber angle	Not quoted
Caster angle	Not quoted
Trail	Not quoted
King pin angle	Not quoted

WHEELS AND SUSPENSION

Tyre size (front and rear)	5.60 x 23 4PR (Stock Moke only)
Wheels	4.50J x 13 (Stock Moke) 5.00J x 13 ROH "Spoker" (Californian)
Suspension	Rubber cone. Fully independent. Front wheel drive

BRAKE SYSTEM

Type	4 wheel hydraulic
Front	Twin leading shoe
Rear	Leading and trailing shoe
Parking brake	Mechanical, acting on both rear wheels.

CAPACITIES

Radiator	3.0 litres (5.25 pints)
Fuel tank	28.4 litres (6.25 galls)
Engine oil	4.83 litres (8.5 pints)

Moke's usefulness for beach fishing, desert wilderness transport, is severly compromised by 2WD.

Opposite: pretty blue Californian is more suited to cobbled roads than bush tracks, although all-road capability of lightweight vehicle surprised our test team.

DOES THE MOKE NEED 4WD FOR OVERLANDING?

With the new 4WD Moke scheduled to go into production later this year, we thought the time ripe to test the 1275cm³ Californian — its 2WD equivalent.

EVERYBODY SHOULD GO Moking at leat once in their lives. For this goggle-eyed, boxy utilitarian vehicle is the epitome of automotive pragmatism.

To many, Alex Issigonis place a blight upon society when he designed the Moke's cousin, the Mini, in the late 1950s. Yet within a few years the east-west engined Mini had shot to the top of small car sales in both England and Australia.

In the mid-60s, the Moke followed. But by 1968 Australia was the only country to continue to realise the vehicle's potential. Now the Moke, made only in Australia, is exported to 63 countries, more than any other locally-made vehicle.

It's not until you clamber (you can't step) into a Moke that you realise how many of the little boxes there are on the road.

With side-curtains removed, you can peer along exhaust pipes and see the valves opening and closing in the diesel engines of concrete movers — that is, if the exhaust clouds don't obscure your vision.

Sydney's nymphettes gaze admiringly at you and hitch-hikers seem to extend their fore-fingers just a fraction further in the hope that you'll allow them to savour the delights of Moking.

Mokes have figured in several *Overlander* travel stories — The Loneliness of

the Long Distance Moker and The Trailer Trail are just two.

Now a 4WD version has been airfreighted to England (on February 7) for head office approval. It's scheduled to go on sale in Australia in August.

We just had to know: how well suited is the Moke to overlanding, and what changes should the engineers make if the double-differ is to be successful?

To find out the answers, we borrowed a blue Moke Californian from Leyland Australia.

The California boasts 5.50 x 13 white spokers, 175 R13 Olympic WT radials, pseudo-denim hood and seats, zip-opening rear hood flap, alloy-spoked steering wheel, locking spare wheel nut, white bull (calf?) bar and rear bumper, dual horns, rubber floor mat and external mirrors. Most important of all, it has a 1275cm^3 engine — identical to the 4WD Moke prototype.

Retail price (plus on-road and pre-delivery costs) is $4370 (*vs.* $3775 for the basic Moke).

Standard features include a large, round speedo (no tripmeter) mounted in the centre of the dash, rocker switches for parking and head lights and windscreen washers, twist switch for two-speed wipers and T-bar choke control. A steering column mounted lever operates the horn, indicators and high beam (no flashers are included). The one creature comfort is a heater/demister.

On the road

With the exception of a shattered windscreen north of Sydney, our highway and round-town Moking proved to be a fun experience.

The standard windscreen is not laminated and our test driver found his vision instantly blotted out. Fortunately, the Moke's open sides provided a quick alternative method of looking to the front.

The Army suggested the Moke should have a laminated screen ten years ago. It's a must on any overlanding vehicle.

Performance from the 'pollutionised' motor — which has the same capacity as the famous Cooper S engines, but is not a Cooper motor — is adequate but not exceptional. The 1275cm^3 engine produced 26.8kW at 4300rpm and 63.5Nm torque at 4035rpm on our dyno test.

However, the Moke's light weight (686kg, or 19 less than the Suzuki LJ80 soft-top) permits excellent performance.

On the highway, cruising at 100km/h is effortless although long distances at this speed are not recommended by its tyre manufacturers, Olympic.

The Moke will accelerate uphill in top gear from 100km/h to a maximum of 130km/h, 5km/h more than a Daihatsu 1600 and 30km/h faster than a Suzuki LJ80.

The ride is remarkably comfortable, considering that the 2108mm wheelbase is 127mm shorter than an SWB Land-Rover). There is no pitching, no doubt thanks to the independent rubber-cone springs and miniscule but functional shock absorbers.

The rack and pinion steering is exceptionally direct. With only 2.3 turns lock-to-lock, it is in stark contrast to the heavy and at times unwieldly steering encountered

Far left, above: Moke engine shares bore/stroke ratio with Cooper S, but pollutionised single carb engine is far cry from Cooper.
Above: Californian package includes front pipebar, spokers, Californian decals.
Left: dash with central-mounted speedo dates from first Mini. Steering is exceptionally light.

on heavier vehicles. There is no steering slack nor vibration, except on rough roads when high speed induces quite violent shuddering.

Engine noise? The classically throaty exhaust note is usually drowned out by other traffic noise; besides, soft-tops were never renowned for their quietness.

The Moke is the most initimate car we have driven. A passenger sits so close to the driver that it is difficult to change gears without fondling his/her legs. Togetherness does make conversation a trifle easier, given the external traffic din.

And it isn't as though you can get away from your passenger. The seats do not recline and can only be moved forward 25mm by undoing four bolts and relocating the seat in a new set of base-frame holes. The seats, are comfortable, but lack lateral support. Seat belts are lap only.

Shifting gears requires you to lean forward; the shift is a short, centrally-mounted floor lever which demands gorilla-like arms.

The small clutch, brake and accelerator pedals are tucked beneath the steering column; anybody wearing Number 10 bushwalking boots could find themselves accelerating and braking simultaneously. The clutch pedal is slightly offset but for the bush, where lightning gearchanges are the norm, ideally the pedals should be spaced further apart.

In the rough

We have been highly sceptical of Moke

SPECIFICATIONS: Moke Californian

Engine
Type: 4-cyl, in-line ohv
Bore x stroke: 70.61 x 81.28mm
Displacement: 1275cm^3
Compression ratio: 8.8:1
Tested max power: 26.8kW at 4300rpm
Tested max torque: 63.5Nm at 4300rpm

Transmission
Type: four-speed, synchro, in unit with engine; front-wheel-drive
Internal ratios: 1st: 3.536; 2nd: 2.218; 3rd: 1.433; 4th 1.000; Rev: 3.545
Diff ratio: 4.267:1
Gearchange: floor mounted shift
Clutch: 235mm diameter single dry plate

Systems
Fuel: mechanical fuel pump feeding single SU HS4 carburettor from 27-litre fuel tank
Lubrication: full pressure internal gear-type pump, full flow filter, 4.8L wet sump
Cooling: closed pressurised radiator with 12-blade nylon fan
Air filtration: dry element
Electrical: 12V/40Ah battery, 28 amp generator

Running gear
175R-13 Olympic WT radial tyres on 5.50 x 13 four-stud ROH white spokers
Brakes: (front) 214mm discs, power boosted; (rear): 178mm drums; (parking) cable activated to rear wheels.

Steering
Type: rack and pinion
Turning circle: 10.2m
Turns lock to lock: 2.3

Suspension
Front: independent, rubber core spring units, double acting shock absorbers
Rear: independent, rubber core spring units and trailing arms, double acting shock absorbers

General
LxWxH: 3232 x 1448 x 1560mm
Kerb mass: 686kg; payload: 228kg
Wheelbase: 2095mm
Track (F & R): 1245/1264mm
Ground clearance: 190mm (under sump guard)
Fuel consumption: 9.5L/100km (around town); 8.3L/100km (on highway)
Range: 325km
Top speed: 130km/h
Price: $4370, plus on road costs (basic Moke: $3775)
Options: fibreglass hardtop ($995); dropside tray (495)
Colours available: red, white, blue, green
Supplier: Leyland Australia, 332 Oxford St, Bondi Junction, NSW 2022.

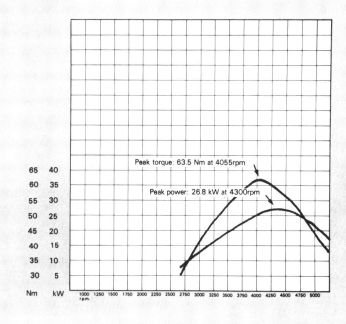

MOKE TEST

owners who claim miraculous feats of derring-do in the bush. Leyland itself does not pretend that the Moke has all-terrain capability, although any company which let Hans Tholstrup drive one from Tasmania to Cape York must feel the vehicle has above average 2WD bush ability.

On rough forestry roads, the Moke shudders on corrugated corners. This is a result of the front-wheel-drive. The excellent on-highway tracking vanishes in the bush unless you use the vehicle's momentum to help you through corners. It's best to rally the vehicle rather than drive it like a normal car.

Loose, rocky climbs are best tackled in reverse, so that the engine weight acts directly through the driving wheels. Otherwise, the front wheels lift and traction is lost. In reverse gear, the Moke conquered one steep hill in five "stabs" that a Daihatsu F20 failed to climb in 2WD (it did the job effortlessly in 4H). With a fair amount of run-up, and some muscle-power on the steering wheel to keep the front wheels biting, the Moke did climb some hills we expected would call for the tow-sling.

We nearly aborted our sand driving test. Even with tyre pressures down to 100kPa (normal inflation pressure is 225kPa) the vehicle bogged down — until we discovered the secret.

Stretches of flat, soft sand should be tackled at at least 50km/h in second gear. This system worked well until the crown became too high or speed dropped for a bend. The Moke has little ability on twisty or hilly sand tracks. We suspect the 4WD version will fare considerably better and may even outperform the Subaru Brumby.

The 190mm ground clearance under the sturdy sump bash plate is equal to many bigger 4WDs. In rocky terrain, this ground clearance was appreciated, especially as the vehicle can clamber over rocks which would damage other 2WD cars.

Suspension travel is limited; it is easy to lift a rear wheel, but the front wheels can pull the vehicle over obstacles which ideally should be tackled in 4WD.

However, rough terrain highlighted another problem which must be overcome in the 4WD version. The SU carburettor was often starved for fuel when the vehicle was on odd angles. The monstrous fuel tank filler cap also did not seal properly and fuel vapour filled the "cabin" as the liquid sloshed around in the pannier-mounted 27L tank.

The Moke measures only 3232mm x 1448mm. With the canopy up, overall height is 1560mm and only 1400mm with the canopy folded. The light steering and good but not exceptional 10.2m turning circle combine to allow the Moke to penetrate narrow bush tracks which might deter larger vehicles.

But it's not all rosy. One major improvement Leyland must make to its 4WD is the vehicle's fording ability. A plastic shield (ex-Mini) may keep water away from the distributor cap but the multi-bladed plastic fan is perilously close to the side-mounted radiator. Undoing the fan belt calls for a long, thin ring spanner and nimble fingers. A quick release mechanism which would allow the alternator to be detached and the fan belt freed would be one solution. A comprehensive owner's manual, à la Land-Rover, would be a good back-up.

The rear Californian bumper should also be re-mounted. It is bolted onto the rear body panel, and any bingle in the back end would damage that panel.

The rear wheel only hand-brake, too, should be modified; it's all right for bitu-

men but on steep bush tracks, the ultra-light vehicle slides downhill with the wheels locked. The front disc brakes (rear drums) spit brake pad lining dust all over the nice spoked white wheels; either the compound is too soft (likely) or the discs are not machined correctly (unlikely). If the compound is too soft, then it will not last long in bush conditions.

No tow hooks are provided: an oversight, in our opinion.

But for overlanding, perhaps the vehicle's biggest drawback is its lack of storage space. There are two "toolboxes", the covers of which are held in place by twist-lock screws which a thief could undo with a 5¢ piece. The hood can also easily be undone (if it couldn't, a razor blade would gain access), and so radios or any other valuable accessory or equipment is there for the taking. A fibreglass hardtop with locking doors is an option, yours for $995.

The canopy is a five-section, vinyl coated fabric held on by lift-the-dot studs. The front side curtains have flaps through which to pay the Bridge toll in wet weather. The rear flap can be unzippered, rulled up and tied in place by two retaining straps. The hood is easily managed by one

Above: because of weight transfer, Moke was better hill-climber in reverse than conventionally. Left: rear bumper would damage body panel in accident.

person; it can be folded flat around the bows or removed and stowed in a vinyl pouch.

Visibility is reasonable: the rear flap and four side flaps have clear, vinyl windows. A centrally-mounted rear vision mirror and two rather flimsy windscreen mounted external mirrors keep you in touch with what is going on behind.

Cleaning the vehicle is simple. Take out the front, tube-rubber floor mat and point the hose. Water and dust will drain out through holes in the floor. Mudguards are easily cleaned, although the rubber shields can harbour mud. Although the boxy body is primed, its acute angles would make rust its greatest enemy.

An extra fuel tank mounted in the right hand pannier would double the vehicle's range (and cut down on stowage space). We achieved 8.3L/100km (34mpg) cruising and 9.5L/100klm (30mpg) around town, which would give a 325km range on the open road, not quite sufficient for long distance touring. A 4WD version would probably use more.

Overall, the Moke is cheap, basic transport. It does have a certain trendy snob appeal, but its nimbleness in city traffic, combined with on-highway performance and economy, make it a far more sensible week-day commuter than a lumbering Land-Rover or Land Cruiser.

The key word for Leyland is IF. The 2WD Moke has certain problems which must be rectified if the vehicle is going to withstand the demands owners of the 4WD version will impose on it. And, of course, price will be an important factor, since without a transfer case it will lack the rough stuff performance of, say, a Suzuki. But that is the type of vehicle which the Moke will have to meet head on. A price tag of around $5500 may do the trick and earn Australia more export income into the bargain.

Almost (And Maybe) The 4x4 MOKE

The Moke gains four wheel drive and is then shelved as all Moke production finishes. But is it all over?

AUSTRALIAN INGENUITY is something to be very proud of, especially in the field of engineering. Our forefathers, in particular those in remote rural areas and on the goldfields, had to improvise perhaps more often than counterparts in other colonies.

Tools and equipment had to be made from whatever they could lay their hands on. Nice new machinery, imported to the antipodes under sail, was extremely expensive, took ages to get here, and 'eons to get to the man on the land. Often, when an item did finally arrive, it was not well suited to our conditions and needed to be extensively modified.

Happily, our inventiveness has not dwindled too much over the years and most farmers can still build a better mousetrap when necessary. Likewise, our engineers can meet most challenges when marketing men and accountants see the need for something new or different.

A perfect example of this is the four wheel drive Mini Moke, conceived and built in Australia. Unfortunately, the intriguing machine looks unlikely to reach production, as the conventional Moke is now all but dead — more is the pity.

Production, which began in 1966, has ceased and, once existing stocks are sold, there will be no more new Mokes anywhere in the world, for Leyland Australia was the last plant to build the box on wheels.

On learning the Moke was doomed, I made arrangements to visit Leyland Australia's Engineering Services operation to investigate the four wheel drive prototype, and to take a last nostalgic look at a vehicle which is very close to my automotive heart.

The first new car I bought was a Moke, in 1966. Actually, it wasn't exactly new but an ex-press test car. Boy, if I only knew then what I know today!

I ended up breaking the poor thing's back when I rolled it at Lakeland Hillclimb but that was after two years of circuit racing it and a previous year of observed section trials, rallies and gymkhanas. Before committing it full time to track work, I used the Moke for its intended purpose, going bush and having fun.

Rather than go into details, which would upset some readers who don't believe I have a responsible attitude to the bush, I'll simply state that, on one occasion I managed to get the Moke stuck in a tree 15 feet off the ground, and on another jaunt into the wilds, about as many feet under salt water in the surf. That vehicle and my youth have many common fond memories!

One of the first things I did to it was fit Hillman Imp wheels, giving it better ground clearance and taller highway gearing. Leyland, or BMC as it was then, followed suit a couple of years later but with 13 and not 12 inch wheels replacing the tiny 10 inch originals.

Another Moke joined the Hicks' stable for a while a couple of years later. We ran that in Rallycross events, until my wife T-boned a spinning car in a ladies event at Calder.

So much for my Mokes. Now to the one I would like to own today.

In the mid-70s, Leyland Australia saw the potential for small four wheel drives, with the Suzuki, Daihatsu and Subaru machines meeting with near instant sales success. The Moke, which, according to the Oxford dictionary, is a slang word for a small Australian donkey, had the image of a cross country vehicle, but not the real physical ability.

Under the care of Ray Habgood, general manager of Leyland Engineering Services, the concept for a four wheel drive Moke was taken through to the prototype stage. There had been earlier all wheel drive Mokes and Minis but these had a complete second

DEEP IN THE STEEL and concrete jungles known as car manufacturing plants, any patient budding Hairy Bottler can find rare species of automobilia never seen in populated areas.

Sometimes, strange skeletons are all that remain to be found of particular species, decomposing to eventually merge into the rugged areas, where old jigs and other rubbish of the jungle piles up, as market climates change.

Grossly distorted beasts often abound, victims of shattering conflicts with barriers. In a cruel ritual, conceived that we may better survive unforseen battles, these beasts have been sacrificed — forced to suicide under the seditious gaze of natives.

If one is careful, and approaches the subject with diplomacy, one can sometimes convince the tribes which inhabit these jungles, to let you see some of these rare beasts.

And, if you are especially tactful, and bribe the tribe's chiefs with promises of trinkets in print, they may even consider allowing you to awaken some of the near extinct monsters, and study them in detail.

A northern tribe, with very different customs and traditions from their larger and more regimented southern native counterparts, agreed to take me on a guided safari through their far flung jungle, near Bankstown in the colony of New South Wales.

Two species, an Ant and a Moke, had long intrigued me. Thus, armed only with a camera and a pen — and a change of jocks and sox — I risked life and limb, leaving the sanctity of the Garden State, or Mexico, as the primitives to the north call it.

On arrival at the jungle I was given a tour of an area known as Leyland Australia's Engineering Services — it was enough to sicken even the most hardened automobilia conservationist.

Here, Frankenstein tortures were being carried out, with dismembered beasts lying everywhere. Under the guise of 'improving the breed' experiments were underway on racks, presses and in cold clinical labs.

Ignorant of the fates which awaited them, many beasts were being prepared for the most horrific treatment. Here, far from the safety of dealer showrooms and consequent care and attention from loving owners, a handful were being given their last rites. These were to be the press test cars.

One shuddered at what was in store for these innocents. But, as is the case in tribal marketing wars, there will always be a toll. Some must die so others may live.

Venturing through the jungle, we came out on a vast plain, covered with beasts of every shape, size and age. It was frightening. Felines took pride of place, with Jaguars sunning themselves, often with Lions nearby, poised on the front of Peugeots.

Under the shadows of huge Terriers, regal beasts were scattered here and there. A sovereign or two, and even an old Princess, could be seen.

In stark contrast to these regals were the lowly ones, a form of donkey — the Moke!

It was a sad sight, row upon row of newborn Mokes patiently awaited transport to market, little knowing they were the last of the breed. Now they are destined to become extinct — cold, calculating genocide has been perpetrated by the tribe's chiefs. It is a crime, for Australia was the Moke's last breeding ground in the world.

For me, it was specially saddening. I had, owned and raced two Mokes in my youth and, while I openly admit to cruelty, I had a very deep affection for them.

Having dissected Mokes and performed several major organ transplants, I felt I would easily recognise the very Mokus Quadrus, which had drawn me to LA's most easterly outpost but it was cunningly disguised and bore no distinguishing marks.

Only by looking up under its rear end could you tell that here was the perfect issue of genetic development.

My other quarry, the Ant, had escaped only a short time before. It had been sold to a native of the tribe, who, subsequently, went walkabout, never to be seen again. A tribal leader told me the Ant was in good hands and would not be harmed, so, while my trek north was to study two rare species, I had to console myself with just seeing the Moke.

motor and transmission added, to make them Twinis. They were expensive, dangerous, and bloody quick in 'S' trim.

The Australian machine had to be realistic and competitive with the little-uns from Japan.

Initial design parameters called for the standard body to remain largely unchanged, engine and transmission alterations had to be made by modifying fully built up power units and costs were to be kept below $880 over the then current 998cc Moke, with tooling and development costs to be amortised over 2000 vehicles per annum for two years.

All the criteria were met. The new vehicle was going to be put into production, until the whole Moke business was shelved in favor of more profitable activities, such as building Peugeots.

Had the four wheel drive gone into production, there was yet another spin-off which Leyland was investigating — a rear engined sports car!

More of that later now. Onto the prototype. Leyland engineers blew the cobwebs off it, charged the battery and let me loose in, under and over it.

Visually, the Quadra-Moke (our name, not theirs) is almost impossible to pick from the mere mortal machine. Only at the rear, where there is a change to the rear crash bar mounting and a very obvious differential, standing out like the proverbial 'dog's', can a difference be seen.

Inside, the story is much the same, only an extra lever mounted on the centre tunnel (which became a legitimate transmission tunnel) gives a clue to the difference between the two and the four wheel drive vehicles. This lever operates a dog clutch, which engages the rear drive.

Had the four wheel drive reached production, it would have been decked out with appropriate badges and cosmetics to herald the doubling of drive capability.

When the mechanical modifications are studied, it appears the Moke was intended to be a four wheel drive, right from the moment the vehicle was first conceived. One wonders how many normal Mini sedans could have been sold with four wheel drive as an option.

Quite simply Leyland engineers added a bevel gear to the side of the existing Mini

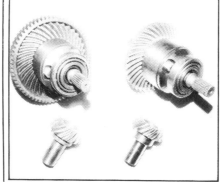

Top: It might look like an ordinary Moke, but note the flange facing rearward from the gearbox. The tailshaft passes above the plate beside the exhaust.

Left: On the left is the front diff, which drives to the front wheels via the outer ring gear, and to the rear via the bevel gear. On the right is the rear diff.

crownwheel, taking the drive out the back of the original diff housing to the new rear axle via a one piece prop shaft. The only significant changes up front, other than the power take off, were the re-routing of the gear change mechanism and the moving of the motor forward about a quarter of an inch in the original subframe. Even the standard transmission guard remained unchanged.

On the way back to the new rear end, things get a little crammed in the relatively small transmission tunnel — which was originally only required to house the exhaust pipe and provide strength to the floorpan. In addition to the exhaust there is the fixed tailshaft and the rod which engages a dog clutch immediately in front of the rear diff.

Save for larger shock absorber mounts on the inner faces of the wheel boxes, the body is unchanged at the rear. A complete new rear subframe has been fabricated using the same system of trailing arms, but with a wide front pivot and some of the rubber cone suspension components borrowed from the front end. New trailing arms, with conventional Mini constant velocity joints were designed. These were to be used on both the two and four wheel drive machines — albeit minus the joints in the normal car.

The rear housing of a conventional Mini diff is used to house the aft diff, with a new cast front half casing onto which the dog clutch mechanism is mounted. The diff' housing, together with the appropriate guard, are mounted on the new subframe.

Axles, joints and many other items were all derived from existing Moke production items, making for relatively economic production and fewer spares needed in parts inventories.

Naturally development of the four wheel drive Moke was not simple and there were a few handicaps which just had to be tolerated. For example the vehicle would have been single range, would have had a less

than ideal ground clearance and smaller than desirable wheel diameter — the last two problems were interelated.

Even so, the final product would have been a worthy competitor for the Subaru four wheel drive station wagon, which at that stage had only a single final drive range plus wheel and clearance restrictions. Skiers, for example, do not need massive ground clearance for snowline driving. There optimum traction is the most important thing.

In all probability the 1275cc power unit would have been used and limited slip differentials front and rear would have featured.

Development of the 4x4 Moke produced quite a few headaches for the Leyland engineers, not the least of which was the overall final gearing for rugged off road work. The tallest diff available 'off the shelf' was 4.26:1. Though acceptable for simple wet and muddy conditions, it was less than desirable for mountaineering.

After several attempts at cutting new crownwheel and pinion combinations, Leyland finally settled on a ratio of 4.33:1. In the meantime they had tried to obtain better ratios by fitting new gears on the step down from the crankshaft to the gearbox. Engineering limitations dictated that these gears would have to be square cut. According to Ray Habgood, you could hear them coming from a couple of miles away! Anyone who has run close ratio square cut gears in a Mini can attest to the noise!

Even after having decided on the diff' ratio, problems kept on coming. Initial testing found the pinion teeth to be too weak, with the internal splining onto the gearbox main-shift too close to tne outer teeth. This was solved by cutting a much longer pinion, which spread the load further onto the splining. But this brought to light more problems. With strong gears, the extra forces involved in optimum traction via two limited slip diffs found a weak spot in the cross webbing of the gearbox casing. As recent Mokes have used an imported gearbox, the strengthening required would have needed to be incorporated in all overseas Mini production -- notably in the Mini Metro recently introduced in the UK to replace the old Mini. For all Minis in the world to have an extra strong gearbox casing just to satisfy the needs of a relatively few all-terrain Mokes seemed to be a folly from the British point of view.

Had the 4x4 Moke, and indeed the Moke itself, been seen as an on-going commercial proposition in Australia, Leyland engineers here would surely have solved the problem by other means. In fact a private Moke owner in South Australia has apparently converted his machine using some Leyland parts, and has achieved the strength by means other than a thicker webb in the casing.

Obviously Leyland engineers felt they could overcome the problem, for in their minds, and on paper to a certain degree, there were other logical variants on the four wheel drive base.

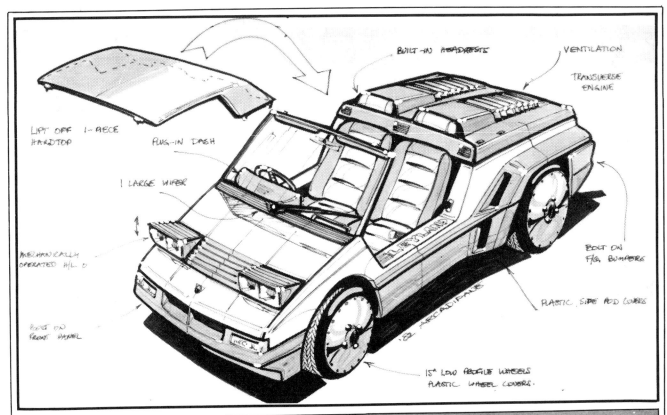

The rear subframe, for example, had been designed so that it would accept a complete Mini powerplant. Thus it would be possible to build a twin engined Mini Moke. Can you imagine two stove hot 1275cc Cooper 'S' motors in a super lightweight Moke body. Wow!! If you really wanted to get mobile, you could add a turbocharger to each donk! The mind boggles!!

It is unlikely that Leyland Australia would have marketed its Twini Moke but it would have been an easy task for an enterprising aftermarket business to offer optional second motors, or, for that matter, for owners to carry out their own installation. What the provision for the second unit did, however, was to make it possible to delete the front motor, and build an instant rear engine sports car.

Leyland put a great deal of thought into this possibility, and had even gone so far as to commission numerous artists' impressions of a rear engine Moke fitted with fibreglass covers front and rear. Tentatively it was called an MG. Had it gone into production, it would have been the first rear engine MG ever built, and also the first such vehicle from any of the BL marques.

The cost would probably have been only a thousand or so dollars more than a conventional Moke, for there were no additional mechanicals, and cosmetic trim items in addition to the fibreglass front and rear covers were not extensive.

The MG and the four wheel drive Moke would surely have found a ready market here, and could well have become 'cult cars' for the trendies, while the Moke would have also been able to fill a very useful role on the land — either for fun or function.

Near the beginning of this article it was mentioned that the four wheel drive looks "unlikely to reach production." "Will not" was avoided, as there is a remote, repeat remote, chance that an aftermarket company may take up where Leyland left off. The company, which wishes to remain anonymous until decisions are made, has had 'meaningful' discussions with Leyland, and is currently involved in other four wheel drive engineering.

In closing I am compelled to pay my own little tribute to the passing of the Moke, by quoting a poem published some years ago in an American motoring magazine — Road & Track I think:

Ugly little Mini Moke,
Are you just a British joke,
People laugh as you fly by,
Like a shithouse in the sky,

TYPE & PRICE

Make	Leyland
Model	4x4 Moke
Code/type	Utility
Manufacturer	Leyland Australia
Seating capacity	Two
Base price	$6000 (estimate)

WEIGHTS & MEASURES

Kerb Weight	632 kg
Wheelbase	2090 mm
Track f/r	1240/1260 mm
Overall	
Length	3230 mm
Width	1450 mm
Height	1600 mm
Clearance Sump	200 mm

ENGINE

Type	Petrol
Cylinders/capacity	4/1275 cc
Bore/stroke	70.6 x 81.3
Compression ratio	9.5:1
Fuel requirement	Super
Carburation/Injection	SU
Fuel pump type	Mechanical
Air cleaner type	Dry paper element
Valve gear & drive	OHV – chain
Ignition type	Coil & points
Block & head materials	Iron
Fan type	Fixed
Maximum power	40 kw at 5,250 rpm
Maximum torque	91 Nm at 2,500 rpm

TRANSMISSION

Gearbox type	4 speed manual
Shift location	Centre tunnel
Clutch	Hydraulic
Gear ratios	
1st	3.526:1
2nd	2.218:1
3rd	1.433:1
4th	1.000:1
Reverse	3.545:1
Transfer box type	N/A
Casing protection/clearance	
Diff ratio	4.333:1
Free wheeling hubs	N/A
PTO facility	N/A

PERFORMANCE

Maximum speed in gears (km/h corrected)	
1st	40
2nd	64
3rd	98
4th	130
5th/overdrive	
Accelleration (high range only)	
0-60 km/h	6.6 seconds
0-80 km/h	9.9 seconds
0-100 km/h	16.0 seconds
70-100 km/h	9.0 seconds
Standing start 400 metres	
Elapsed time	20.1 seconds
Terminal speed	109 km/h

Portuguese Production Mokes

The Australian Moke production line stopped in 1982. But the Moke was not yet dead, and after a gap of two years production started up again, this time in Portugal. Initially production was of the same Californian Model produced in Australia. Although Austin Rover sold these through its dealers around the world they decided not to sell them in the UK. This job was taken over by component car manufacturer, Dutton. Dutton re-launched the Moke with a blaze of publicity, and it was welcomed back with open arms by the press. However, in the end difficulties arose at the Portuguese factory, and at the UK end and few ever arrived.

The difficulties at the Portuguese factory turned out to be serious, and Austin Rover sent in a consultant. It was through his efforts that Moke production still continues today. Jim Lambert (pictured below with the 1988 Moke) took the production to a new factory, revitalised the quality, brought the Moke back to profitable production, and ran the operation until the end of it's British ownership. The excellent article from Autocar *A man and a Moke* tells the whole story. The Moke survived another threat of closure in 1987, but in 1989 the end came, with the last Austin Mini Moke produced in July of that year.

The changes made by Jim Lambert developed into the new 1986 Moke, with a return to more standard mini parts. The emphasis changed from the rugged jeep look, to the fun-loving, sun-seekers transport, and this proved very popular with demand outstripping production. In particular it was taken on as a holiday hire car in many parts of Spain, France and the Carribean - a few even found their way back to the UK. The Moke continued to be developed, as you can read in the article by Mini Moke Club member Peter Jones; the production line a far cry from the original at Longbridge 25 years earlier, but the quality was excellent and the price - a little over 12 times that of the first Moke.

photo by Peter Jones

Moke Californian

TESTED

One of the original fun cars of the swinging sixties, the revised Moke Californian provides good value, great fun and plenty of attention wherever it goes.

For those too young to remember, it was two decades ago, during the early sixties that the Mini Moke was first produced by BLMC. Originally intended for military purposes only, a batch were sent to the Fighting Vehicle Research and Development Establishment at Chobham for evaluation as basic forms of transport for personnel. The Moke's advantages were obvious; sturdy enough to withstand a parachute drop, light enough to be lifted by helicopter and compact enough to be stored on top of one another, but in the end it was decided that ground clearance was insufficient. Rather than ditch the whole project, BL decided to test public reaction to the Moke; after all there had been many weird and wonderful Mini conversions already by that time, albeit produced only in small numbers by specialist firms.

From the moment of its introduction in August 1964, it became obvious that this marketing strategy was correct; the public loved it. It became just as fashionable to be seen in a Moke as it had years earlier with the original Mini saloon and it wasn't long before several coachbuilders realised its potential and produced modified versions. One that immediately springs to mind is the Surrey Moke from Crayford which sported a striped and fringed canopy and spare wheel cover. Indeed, this is the car which appears in the re-run of the popular TV series, 'The Prisoner', starring Patrick McGoohan.

The Moke also turned out to be surprisingly competitive in motorsport. Hans Tholstrup tackled the London to Sydney Rally in 1977 and four top class drivers, including Bernard Cahier, entered the gruelling Targa Florio in 1963. This admittedly was in a Twini — a derivative created by the Mini's designer, Issigonis, which featured an engine at either end. It was in one of these machines in fact, that John Cooper fractured his skull, having somersaulted end over end from a speed of 100 mph!

The death of the Moke in this country came about in 1968 when Customs and Excise, who for four years had classed it as a commercial vehicle exempt from Purchase Tax, suddenly changed their minds. BLMC obviously thought that a drastic drop in demand would result and transferred the whole operation to Australia. Over a decade of production in that country was due mainly to the Moke's popularity as a hire car for tourists in the Caribbean and other fashionable resorts. There was one practical drawback with this situation though — replacement parts were sent from the UK, and any warranty claims on engine parts were subject to long delays as paperwork flew halfway around the world. The costs involved finally became exorbitant and production was switched to British Leyland Portugal Automoveis at the beginning of 1983. Austin-Rover over here felt that there might again be a market and a special place in the memories of enough people in this country to make it worthwhile for importation. They did not feel, however, that the Moke would sell in sufficient numbers to justify offering it through their own 1500-odd dealer network, so offers were invited from other parties for the franchise. From the half dozen contenders, some of whom were large Austin-Rover dealers, the well-known entrepreneur from Worthing, Tim Dutton, won the day.

Already a household name for their competitively priced component vehicles, this is the first complete, production car ever to be seen in their showroom. The 998cc Mini engines are still exported from this country to Portugal where, once installed in the Moke, the complete cars are shipped back here at a very small cost.

Interior is functional with rubber slat matting instead of carpet and very obvious heater hoses. Driving position is very comfortable.

STYLING, ENGINEERING ■ ■ ■

Because it was originally intended to fulfill a military role in life, the Moke has no pretentious or unnecessary styling features. Function was the name of the game and rather like the Land Rover, which has changed hardly at all for many years, the Moke is specifically orientated towards utility The barrel-like bonnet, peculiarly shaped radiator grille and practically upright windscreen combine to make it about as aerodynamic as a brick wall, and on closer inspection, the Moke turns out to be a curious mixture of straight lines and sharp corners contrasting with that rounded front end, Mini headlamps and distinctive wheelarches. Whatever your opinion of the Moke, it has to be admitted that it has a certain indefinable appeal, in rather the same way that some of the Japanese 4X4s, such as the Suzuki SJ410, do. One enormous improvement in both styling and engineering terms is the introduction of 13 inch diameter wheels. When looking back at photographs of the Moke from the sixties, one is immediately struck by how ludicrously small the original 10 inch wheels were.

In practical terms, the Moke is excellent. Its space-saving design ideas include location of the battery and tools in the driver's side sill, whilst the eight gallon fuel tank lays in the same position on the passenger's side. Behind the rear seats there is a long, narrow luggage compartment, covered by two sliding doors which are lockable, so that valuables can be safely left in the car. Whether or not the car itself will be safe is another matter — apart from the usual steering lock, there is no form of security.

The compact but rather tall A-series engine as used in the Mini and Metro fits snugly into the rotund engine bay and will never meet with puzzled looks from garage mechanics; its reliability and long life are well proven and servicing should be well within the capability of those who like to maintain their own cars. This is made much easier by the fact that the engine cover can be removed completely by undoing the rubber toggles and lifting it out of its hinges.

The little 998cc unit produces just 40 bhp at a fairly low 5,200 rpm and 50 lb ft of torque at 2,500 rpm, coupled to the Mini four-speed gearbox. Suspension is independent all round by unequal length transverse arms, tie rods, springs and dampers at the front and trailing arms, springs and dampers at the rear. Steering, of course, is rack and pinion and drum brakes all round cope very well with bringing the nearly 13 cwt Moke to a halt.

DRIVER COMFORT ■ ■ ■

Access to the driver's seat is fairly straightforward, requiring the front window to be partially unzipped. If the hood is down, great care is needed to avoid getting skirts or trouser legs dirty on the quite wide sills. Once inside, the front window is easily zipped up again and the Moke feels quite snug; a rather similar feeling to zipping up a sleeping bag or tent!

The high-backed seats are very comfortable and provide ample support for most frames, although two large people might find it a rather tight fit. This is because a good foot of the passenger compartment either side is taken up by those sills, and consequently the seats are rather close together, definitely ensuring that you get to know your front seat companion round tight left hand bends! Instrumentation is sparse, comprising the huge Mini speedometer mounted centrally in the dashboard and surrounded by switches for lights, foglamps and any auxiliaries fitted. Stalks deal with indicators and horn on the left, wash and wipe on the right and apart from a guel gauge, that's just about it. There's the choke knob of course and heater controls — a pull out knob for heat, a slider marked 'car', 'screen', 'off' and a fan on/off switch — which although basic is surprisingly effective. The only gripe our testers could find about the driving position concerned the handbrake. This was

Compact little A series has all routine service items accessible and should present no problems at all for keen DIY owners.

set too far forward and meant bending forward to apply it, and although this was quite easy due to the fitment of inertia reel seatbelts, it **was** irritating.

All round visibility was better than expected with the hood up, thanks to the fact that the canvas does not extend right up to the windscreen edge. Instead, there is a long glass quarter light, so that the large wing mirrors are not obscured. We did suspect however, that in wet weather the wiper blades would be too short to clear a large enough area of windscreen for good forward vision.

On first getting into the Moke, it seems as though the steering column is skewiff, but this isn't noticed when on the move, and the gearlever could ideally have been a few inches longer to make first and third easier to reach.

IN ACTION ■ ■ ■
There is no point in even pretending that the Moke could be an everyday car for the high mileage user. As we found out on the return trip to Worthing to hand back the car, it definitely isn't much fun on a motorway or at least not in chilly November! The driver suffers from considerable buffeting with the hood down, and is subjected to a lot of noise with it up. Where the Moke excels is cruising (or should we say posing), around town, on A and B roads or in the rough.

The Mini's handling is legendary and the Moke, with its improved damping, is even better. It sticks, leech-like, to the road when chucked into corners, and although there is some tendency to understeer as with all front wheel drive cars, it can be easily corrected.

The Moke's off road ability never ceased to amaze our testers. The eight inches of ground clearance meant that it could be hustled across fields, over rutted, stoney tracks and even driven through water splashes as our photography confirms.

With only 40 bhp, a burn-off at the lights, even against a Fiat 127, is out of the question. Although the gearchange is very slick, the tall gearing of the A-series and primitive engine mountings mean that running each gear to the limit is not a very pleasant experience. Far better to get into top as quickly as possible to minimise noise levels and then work up to a top speed of around an indicated 87 mph. Any slight incline or even a sustained gust of wind can knock as much as 10 mph off this speed however and it's very much a case of boot to the floor on motorways. For A-road overtaking, it's a toss-up between getting a good speed up in fourth, coming in close for a look and then venturing around the obstacle, or suffering the'noise which a downchange to third entails. Although third **is** accelerative enough for overtaking, it really becomes impractical at anything over 45 mph, because of these noise levels and mechanical cruelty to the engine.

The gearbox was very nice in action, with a very short clutch travel and matching short throws between the gears, although a badly adjusted clutch in our test car meant quite a lot of effort was needed to engage first and second gears. Another characteristic, we suspect peculiar to our Californian, was a tendency for the gearlever to jump out of fourth when backing off the power.

THE DAILY MOKE ■ ■ ■
Despite some of the criticisms we have levied against the Moke, it turned out to be a surprisingly practical little machine for everyday use — one of our testers really fell for the Moke and tried to persuade Dutton that a 12 month long term test would be a good idea!

Most impressive in practical terms was the hood design which could be tailored to suit weather conditions or the mood of the driver very quickly. For really cold days obviously all the panels are left in place, but a sudden rise in temperature or the need for some ventilation means simply stopping to unzip the plastic rear window section or taking out one or both rear side panels. This ensures good throughflow but no draughts to

Standard fitments include wing mirrors, mud flaps and nudge bars front and rear to make Californian great value for money.

the front of the car. To put the hood down completely is a five minute operation and easily managed by one person. All the windows are unzipped and the stud fastenings pulled to release the bottom edge and the panels can then be stowed in the lockable compartment at the rear or on the floor behind the front seats. The stud fasteners are then undone along the top of the windscreen frame and the whole assembly of canvas and aluminium frame can be neatly concertina'd back behind the rear seats. One design feature that was particularly appreciated was the provision of a horizontal flap in the front doors, enabling front seat passengers to put their hand out and undo the holding toggle, prior to unzipping the door from the inside.

On the minus side is the petrol filling procedure. The filler cap is set into the passenger side sill which is only about 10 inches above the ground, so the driver really needs to squat rather than just bend down, to get to it. Upon putting in the pump nozzle, one immediately notices an amount of resistance; the filler will only go in about 2-3 inches, due to the odd shape of the neck fo the tank. The problems start when the trigger is squeezed to give full flow; one tester got covered in petrol due to the vicious blowback. Because the nozzle will not go in very far, the petrol has a nasty tendency to flow straight back out of the tank. It's fortunate that the Moke is unbelievably economical (we averaged over 40 mpg all the time), because brimming a nearly empty tank took an average time of 15 minutes.

One enormous advantage the Moke holds over other specialist cars produced in small quantities, is that any problems occuring during the 12 month/12,000 mile warranty period can be dealt with at any Austin-Rover dealer in the country as well as servicing, for those who don't feel like tackling it themselves. Dutton Cars will also be holding a stock of any spare parts peculiar to the Moke.

CONCLUSION ■ ■ ■

To sum up, the Moke's few drawbacks of less than ideally placed handbrake, engine noise levels and awkward petrol tank are more than offset by its tremendous character. This, after all, is what the Moke is all about. For just

Rear luggage compartment has lockable sliding doors and can be used to store door panels or small personal items.

£4,360 on the road, it provides truly alternative motoring and is equally suited to pottering around town every day or for use on a farm. There are many items included in the standard specification that would be optional extras on other, similarly priced vehicles, such as nudge bars, high-backed seats, foglamps and mudguards, not to mention the wheels, hood and upholstery finished in white for no extra cost.

Whenever the Moke was left in a public place, we found a crowd of people around it; ranging from youngsters who had never before seen one in the flesh to middle-aged enthusiasts who commented that they "hadn't seen one of these for years", and judging from the very favourable reactions that we encountered, Dutton Cars are very soon going to be on the receiving end of plenty of orders. The first order to be received, by the way, was from the Onassis family who want four identical Mokes!

The gap of 15 years since the last time a new Moke was seen on these shores is not apparent in its design, and although it could never be called ultra-modern or refined, it is definitely trendy. We cannot emphasise enough how much fun the Moke was and confidently predict it will become **the** car to be seen in during 1984.

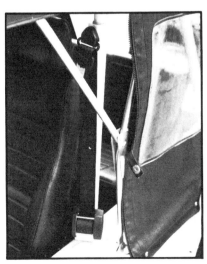

Inertia reel seatbelts are cleverly mounted to the roll over bar. High backed seats are standard.

At a glance...

Body/chassis: Monocoque, all steel. Anti-rust treated.
Suspension: Front — independent via unequal length transverse arms, tie-rods, rubber springs and telescopic dampers.
Rear — independent via trailing arms, rubber springs and telescopic dampers.
Brakes: Drums all round. Parking brake by mechanical drum brake action on rear wheels.
Steering: Rack and pinion
Engine: Four cylinder in line
Cast iron block and head
Bore/stroke 71/81 mm
Capacity 998cc
Compression ratio — 8.1:1
Ignition — contact breaker
Carburettor — single SU
Max power — 40 bhp @ 5200 rpm
Max torque — 50 lb/ft @ 2500 rpm
Weight: 1428 lb/649 kg
Fuel capacity: 8 gals
Performance: Max speed — 84 mph
0-60 mph — 27.9 sec
Price: £4100 (inc car tax and VAT)
Contact: Dutton Cars Limited, (T-AC),
53 Broadwater Street West,
Worthing, West Sussex
BN14 9BY
Tel: Worthing 213421

AUTO FUN

Go with Mokes!

That ultimate Sixties cult car, the Mini Moke is becoming trendy transport in the image-conscious Eighties. A chic modern version, built in Portugal, is now available here and is a lot of fun, as Andy and Nick Hibberd found out

AS A CAR it is hard to take seriously, but the once much-loved Mini Moke looks set to make a major comeback. The popular Sixties cult car was all but killed off by British Leyland in the early Seventies.

Several foreign factories did continue to build it under licence but it was not imported to the UK.

A handful of Australian models found their way to Britain but these tended to be built to withstand the rigours of the Australian outback and were functional rather than attractive.

In Portugal the Mini Moke has progressed with the times and is now a trendy and attractive car.

And, thanks to the entrepreneurial instincts of a car dealer at Bagshot in Surrey, it is on sale in the UK again.

David John, of Duncan Hamilton Ltd, was taking a short cut across the Austin Rover stand at the Geneva Motor Show when he literally fell over a Moke on display.

Designed for fun

IT WAS love at first sight.

He ordered ten on the spot and has been selling about a dozen a month since then.

One of the car's main attractions is that it looks just as at home parked on a beach or outside a top London hotel. And it costs less than £5,000.

But under the bonnet this frivolous little car has not advanced with the years and it certainly isn't quick.

It would probably not be able to keep up with go-faster stripes even if you could find enough metalwork on which to mount them.

Power is provided by an ordinary 998cc Mini engine producing 40bhp and driving the front wheels via a four speed gearbox.

The Moke is not as aerodynamic as an ordinary Mini. The flat front and large windscreen make it about as streamlined as a brick and even a gentle headwind can make a dramatic difference to its sustained 65mph top speed.

But speed is not what this car is all about. The Mini Moke is designed for fun ... with a capital F.

When the sun shines you can take the roof right off and not be encumbered by doors and windows.

When the weather is indecisive you can follow suit by travelling either with the sides rolled up or removed completely but still with some protection against sudden showers.

And when the weather closes in the car can also be closed in, providing protection for all its occupants.

Zippy performer

BUT WHICHEVER way you chose to travel, the car is noisy and there is a lot of wind noise above 35mph.

With the roof completely off the wind whistles past the windscreen and buffets the occupants from every conceivable, and inconceivable, angle.

Climbing in and out of the Moke is also an interesting experience.

There are no doors, just zipped sections of plastic. The zips tend to be hard work to open and close but, having overcome that hazard, you then have to climb over the wide side panel of the car. Unless you are an exhibitionist this is best avoided when wearing a short skirt or baggy shorts.

The Moke's simple heater, the fan is either on or off, is surprisingly effective. But like just about everything else on this car it is quite noisy. At speed, however, it is drowned out comple[tely] by the general cacophony.

In the battle for sound supremac[y] optional radio also loses out.

A tight squeeze

INTERIOR SPACE in the Moke is row, so narrow in fact that the f seats actually touch in the middle do thighs unless a concerted eff[ort] made to avoid contact. But the c[ar is] not uncomfortable to drive.

The light body and simple ru[bber] cone suspension combine to make ride quite bouncy but the seats are enough to absorb some of the ro[ad]

BIG FUN: The Mini Moke has definitely not been designed as serious everyday transport

BEACH BUGGY: The Mini Moke looks just as at home on the beach as it would parked o[utside]

MODERN MOKE: The latest model is sty[lish]

Autofacts

Make: Austin Rover
Model: Mini Moke
Price: £4,995 on the road
Engine: 4 cylinder in line, transverse mounted
Capacity: 998cc
Power: 40bhp
Fuel: 2 star.
Compression ratio: 8.3:1
Valve gear: Overhead valve
Drive: to front wheels
Gearbox: 4 speed manual
Gearing: 16.8mph/1,000rpm
Steering: Rack and pinion
Suspension: Front, double wishbone, rubber cone springs, telescopic shock absorbers
Rear, trailing arm rubber cone springs, telescopic shock absorbers
Brakes: Discs front, drums rear
Tyres: 145/70 SR 13

london hotel. That is just one of the things that have made the car so appealing Photographs: COLIN POOLE

BAD WEATHER FRIEND: The effective hood

As in the early Minis, instrumentation is minimal. The speedometer is large and easy to read but it is the facia's only dial and is mounted in the centre of the dashboard. It contains warning lights for oil pressure, main beam, ignition charge and direction indicators. There is also a small fuel gauge.

There is no glove box in the Moke but there is a small shelf in front of both the driver and the front seat passenger.

Lockable boot

UNLIKE OLDER models, the trendy Portuguese Moke does have a boot. This is little more than a lockable metal box.

The spare wheel is mounted between the rear bull-bar and the car itself. Both the bull-bars are standard fittings.

The battery, jack and wheelbrace are stored in side compartments.

Despite its low sides and vulnerable look the Moke should be quite a safe car to drive. Passengers are protected by a very solid roll over/safety cage.

And the Moke must be one of the easiest cars in the world to park.

Find a tight space, drive the nose in and then a person of average strength would be able to drag the back of the car into position, it is so light.

Collector's item

TO TAKE the Mini Moke any more seriously than that would be to mistreat it. It is purely an simply a fun car.

But if you want one you'll have to move quickly. The Portuguese factory is due for closure on July 31.

Duncan Hamilton's has more than 100 on order but after that the car's future is uncertain.

That means that any Moke bought now could well become a collector's item in the very near future.

● The Moke is imported by Duncan Hamilton & Co Ltd 0276 71010

Auto findings

Maximum speeds (mph)
1st gear	23.2
2nd gear	42.7
3rd gear	65.2
4th gear	60.4

Acceleration (seconds)
0-60 through gears	29.1
40-60 in fourth	23.2

Brakes
Stop distance from 30mph (ft)	35.0
Handbrake efficiency (%g)	32.7

Fuel consumption (mpg)
Test overall	42.0

Government figures are not specified for the Mini Moke as the vehicle is available on personal import only.

Dimensions and weight
Overall length (in)	127.2
Overall width (in)	56.7
Overall Height (in)	57.5
Wheelbase (in)	80.2
Front headroom (in)	37.0
Front legroom (in)	38.8
Rear headroom (in)	34.8
Rear kneeroom (in)	23.0
Turning circle dia (ft)	30.8
Fuel tank capacity (gall)	8.6
Kerb weight (cwt)	14.2

Other data
Speedometer reading at 60mph	64.8

Equipment
Carpets	S
Cigar lighter	N
Clock	N
Folding rear seat	S
Head restraints (front)	S
Intermittent screen wipe	N
Locking fuel cap	S
Radio	O
Trip distance recorder	N

Service and warranty
Major service (miles/hrs)	12,000/3.5
Warranty (months/miles)	12/12,000

ess. They are also quite supportive.

There is a positive side to the hard ide, however. Despite its standard width tyres the Moke corners superbly with virtually no body roll.

It does occasionally kick up a back wheel if cornered too hard but this is ontrollable by backing off the gas or asing the line of the corner, both of which bring an instant response.

Forward gear lever

THE LAYOUT of the car's controls is not too good. The gear lever is set a long way forwards and requires a long tretch to use it.

The handbrake is also set low and orwards, so the driver has to be prepared to have the seat back set at a far more upright angle than would perhaps be necessary with an ordinary car.

By contrast the steering wheel is quite high and lies fairly flat, resulting in a push/pull action for steering as opposed to an up/down movement.

The pedals all have short travel but are firm to the touch.

Unfortunately, the brakes are not servo-assisted and have to be pressed very firmly to stop the car quickly. Some servo-assistance would make them more effective.

Cult status

THE MINI Moke became fashionable in the late Sixties after featuring in the cult TV series The Prisoner, starring Patrick McGoohan.

The Prisoner's complex plot centres on a secret service agent who, without explanation, suddenly resigns his job.

Anonymous forces gas him in his London home and he wakes up in a strange but beautiful village – a village which could be anywhere.

Number Six

HE IS now 'The Prisoner'– known only as 'Number Six'– and he begins a long struggle to find out where he is, who has abducted him and why.

In the series Mini Mokes provide a 'taxi service', taking the villagers anywhere they want to go within the village... but never beyond.

The Prisoner was filmed on location at Portmeirion, North Wales, and MGM's Borehamwood studios.

I AM NOT A NUMBER: The car, props and Portmeirion setting will be immediately familiar to fans of the Sixties TV series The Prisoner

A man and a Moke

Thanks to the efforts of Jim Lambert, the Mini Moke is alive and well in Portugal, at least for the time being. Unfortunately next year may finally see the end of the line. Jon Pressnell reports

In 1984 Jim Lambert thought he had ended his 41-year career in the motor industry with his last job, overseeing the end of the Mini Moke production in Australia where it had become uneconomic.

Before he retired, however, he suggested that Moke production could be transferred to Portugal where there was spare capacity. His suggestion earned him an unexpected new job after just a few months of retirement. Moke production at the Industria Montagem Automoveis factory in Setubal was in a mess; would he mind sorting it out, on a consultancy basis?

"I fell in love with the project... it encompassed everything I'd ever done. I was asked to come here and sort out all aspects of design, production, quality, cost, finance, to get things on a profitability basis. Things were even worse than I anticipated, and the plant in which we were producing the units was going from bad to worse. Labour relations and lack of payment, a disease which was sweeping Portugal at this time, culminated in December 1984, when the plant was declared bankrupt.

"UK, who were really very enthusiastic about Moke because it was going to provide extra quota for bringing in AR products and supporting the export drive became very disillusioned; instead of being a vehicle which would be helping, it was causing them lots of headaches.

"At this juncture we had 260 units sitting in England in boxes; there were another 30-40 units in various states of assembly in the plant, which was surrounded by militant workers."

Lambert had already seen the writing on the wall and had researched five alternative assembly plants. Most appealing of these plants was one at Vendas Novas, 100 miles from Lisbon and formerly an assembly plant for MAN trucks and BMWs.

He put together what he termed a 'salvage project', whereby those units in England and in the factory at Setubal would be assembled at Vendas Novas. "So, on one very wet and windy day and night we absconded

One of the *most obvious differences with Lambert's modified Portuguese Moke is the roll cage and top arrangement. Engine is low-compression 998 A-Plus and driving position typically Mini*

from the plant with trucks and bits and pieces, like thieves in the night, and went to the new plant. In two days we removed everything."

With a staff of three people, Lambert set up the jigs and started training labour and two months later the first Moke rolled off the line.

"Once we'd established that we could produce vehicles, because no-one in England really believed that I could in fact take all these parts into another plant and assemble them, we produced better quality, without any labour problems, and we were also producing at lower cost."

The quality of the Moke is much better than it ever was, claims Lambert. "We receive a technical and quality audit by UK-based personnel. We've had several audits, and each has been better than the preceding one... the last one was quite glowing."

Encouraged by the success of the new operation, Lambert began to think in tems of continuing Moke production after all the salvaged ex-IMA kits had been completed. There was one major obstacle, though: the Moke was being built to the specification evolved in Australia, with a number of costly special or modified parts.

The Australian move from 10ins to 13ins wheels was mainly to blame, necessitating special trailing arms for the rear suspension and alterations to the steering rack, as well as a revised shape for the rear mud-guards. A further drain on profitability was the differential, which had a ratio particular to the Moke and was supplied in penny-packets to Portugal.

Lambert put together a package based on standard Mini parts — low-compression 998cc engine, economy-ratio gearbox, standard 3.4 to 1 differential and, above all, 12ins wheels — which meant not only normal Mini trailing arms but also, for the first time on a Moke, disc brakes. With such a specification, subframe assemblies and wheel/type sets could be shipped ready-assembled to Portugal, further saving costs.

In August 1985 Jim Lambert received the go-ahead: total design and engineering responsibility lay with him, but Austin Rover in Britain would look after homologation and type approval, as well as sales and marketing. "I've always had very good support from England," Lambert comments. "They've been bloody marvellous." ▶

◀ With the assurance of this support, Lambert was able to begin work on improving the Moke. "I never did like it when I visited Australia, and I told the Australians so. They didn't want to know."

The smaller wheels enabled the Moke's rear mudguards to regain their original shape, so Lambert applied himself to the question of the hood — crucial on a vehicle as open to the elements as the Moke.

He's particularly proud of the end result, which is very much his own work (down even to the specification of plastic used for the window panels) and which reflects his canny approach to design: money has been saved, and access to the rear facilitated by eliminating the obstructive hood frame and adapting the roll-cage to provide anchorage for the hood.

The first genuinely Portuguese-specification cars rolled off the Vendas Novas line in February 1986 — with a hefty element (around 50 per cent by parts value) of local Portuguese content.

Hoods, interior trim, roll-cages and bumpers are all made in the same factory in Oporto as the body pressings and the petrol tank, and the plant now makes a tidy profit for Austin Rover Portugal, which has a 50 per cent share in the concern.

Moke production is now a much more tightly run affair and production figures have reflected this new-found efficiency. In the first five months of 1986 more Mokes — 537, in total — were produced than in any of the preceding years.

In world terms such numbers are mere chicken-feed. So what sort of market does Lambert envisage for the car?

"I believe it's the right vehicle for the islands of the Caribbean, for places like Portugal, for Spain, for Greece, for Malta, for the holiday market and for the rich enthusiasts who can afford a Jaguar but would rather run around in a Moke. Everyone runs around in Jaguars, but not many people run around in Mokes . . .

"It's becoming more and more of a cult vehicle these days, as a third car in the garage for the summer . . . It's a holiday-home vehicle. It's also expected to be robust and reliable with high levels of fit and finish."

In practice, the market for the '86 Moke hasn't quite worked out as expected — not that Lambert is complaining! While islands such as the Seychelles, St Kitts and New Caledonia have taken respectable numbers of vehicles, it's been the French who have taken the lion's share, having doubled their initial order of 384 units.

Italy and Spain have proved good customers, too, but Lambert impishly suggests that the vehicle's success in the tiny Caribbean island of St Barthélémy is more impressive: here the Moke has 80 per cent of the car market . . .

What about developing the Moke, then? Would it be worth considering four-wheel drive, for example, to give the Moke cross-country ability to match its mini-Jeep looks?

"I can't afford anything other than what I've got at the moment. It's marginal now for profit. I have to look at every *escudo* I spend. There're lots of things I'd like to do; what I'm trying to do is to improve the aesthetic appeal of the vehicle. Once I start developing 4wd and putting it on bigger wheels, I start making it very unique and very expensive and I don't know how many people would want to buy it. I don't see myself as a competitor to Suzuki, Daihatsu and those other Japanese 4wd manufacturers."

A more immediate issue is that the Portuguese import quota system is to be dismantled at the end of 1987: Austin Rover will thus be able freely to import its cars into Portugal and so the Moke will no longer be of use to the company.

Consequently Austin Rover in Britain has at present only given its support to Lambert's Moke project until the end of 1987. So the future of Moke after this date will be very much in the balance and will depend entirely upon its ability to stand on its own profitability — and on the availability, of course, of Mini power units and Mini running gear.

Lambert has sold 100 Mokes in Portugal in 1986, "just to support the market". With the quota system, that's meant that he has been forced to cut back on Metro imports by a hundred. For each Moke sold in its home country, then, he's sacrificed, on paper, a profit of £2000.

"I believe that as the vehicle is ▶

Jim Lambert and his Portuguese pride and joy

The welding shop at Vendas Novas awaits the Moke

The decidedly unautomated Moke production line

Paint finish is excellent despite rather limited facilities

made here, we should support this market."

All this, however, depends on the Austin Rover board in Britain sanctioning continuation of Moke production. The Moke is profitable, it has excellent sales potential, it gives a sense of pride to Austin Rover Portugal — but is that enough? In February 1987 Lambert will present to the Austin Rover board a paper on Moke economics. On the reception of that paper depends the future of the Moke. ■

PORTUGAL'S MOKE PERFORMS

The Mini Moke has certainly come a long way since it was launched in October 1964, following its rejection by the armed forces for whom it had originally been designed as a pack-flat air-portable.

As simple and utilitarian as ever, it has lost none of its character yet has become much more civilised. The Moke — civilised? In its new Portuguese specification, yes.

The smaller wheels, with their flush Mini trims and ordinary road tyres, are largely responsible. The stance is less aggressive, and the wheels no longer seem to be bursting out of their wheelarches; the straightened-out rear arches and the neat new moulded mud-skirts help, while the new central position for the exposed spare wheel gives the Moke a less military look.

Access has been improved too, particularly at the back, where the rear-most uprights of the roll-cage are now at the very tail of the car, rather than at the front of the wheelarch. Having gained a crosspiece, the cage is beefier, too, and can now be used as a rack for carrying surfboards and the like.

The high-backed seats, Portuguese-designed, give good support, but the driving position is still pure Mini, so won't be to everybody's taste.

The new hood is measurably superior; not only is it made in better quality materials, but it features a three-stage facility, whereby as well as being totally removable, it can either be half rolled back or fully rolled back — and the rear window can be rolled up. Particularly impressive is the effective method of tensioning the fabric, using webbing straps, nylon buckles and over-centre catches.

There's only one snag: because the hood lacks its own frame, it has to be stowed separately, either being put in a hood bag and kept behind the seats or left at home.

Road behaviour is as one would expect, given the Moke's standard Mini running gear. It might be fashionable to sneer at the A-series engine, but it's a willing little unit and it's now a lot smoother than in the past. High gearing in this application means that there's not a lot of 'go' in top, making it often a bit of a struggle to get above 60mph without a down-change, but performance in third is quite sprightly and gives easy overtaking power.

With the aerodynamics of a flying bedstead, the maximum speed of around 80mph takes some time to attain; 60mph comes up in 22.7

... never a squeak or a rattle from the structure, even on the worst of Portuguese roads

seconds, according to a Portuguese road-test. More relevant is the contented way in which the Moke cruises at 50-55mph — albeit with a massive amount of wind noise doing its best to drown out the uninsulated din from the engine.

To those whose Mini-owning days are now a distant memory, it's a refreshing surprise to realise how the Mini's drivetrain has improved.

As for the brakes, non-servoed, they are vastly superior to the old Mini drums. In common with the gear lever, for which a more rearward location is planned, the handbrake is a good reach away.

It's the legendary Mini handling which really puts a smile on one's face. What can you say, when it's all been said so often? Flat, roll-free cornering, lift-off tuck-in to temper the understeer, that lovely, precise, well-weighted steering, with only 2½ or so turns lock to lock . . . bliss!

The Moke really does feel like a little go-kart — even more so than an ordinary Mini, thanks to the minimal bodywork. And all the time there's never a squeak or a rattle from the structure, even on the worst of Portuguese roads.

For sheer personality, there's little to beat the Moke, unless you pay a lot more money, and on a hot summer's day there can be few more agreeable vehicles.

As everyday transport in this country, however, it's simply not a practical proposition — a Moke in a winter rainstorm must come close to the ultimate in four-wheeled misery.

That's not to say that it wouldn't have a limited appeal if it were to be sold in Britain (which is unlikely), not least because of its excellent quality: some, indeed, would judge it better-built and better-finished than today's products from Longbridge and Cowley. What piffle to say that such an endearing and original little car would detract from Austin Rover's current image! ■

The many refinements made to the Moke by Jim Lambert and his crew make the Portuguese version very attractive

The Moke Factory - How a Moke is Born
by Peter Jones

This spring my wife and I again returned to Portugal where we enjoyed a splendid week as guests of Jim and Salome Lambert. During the week Jim (who is in charge of the Moke Factory) very kindly took me to see the Moke production line at Vendas Novas, a 2 hour drive from Jim's home.

A new office suite has recently been completed on one side of the assembly plant, and this enables stock control, reordering, laboratory and development work to be run efficiently on site.

I was fascinated to follow the complete assembly process. At one end of the plant are the pressed body panels which are manufactured in northern Portugal. These panels are fed to the welding shop for the first stage of the body assembly. After body assembly the body and ancillary parts are degreased and go through a newly introduced zinc Electro galvanisation and preservation treatment. The body and component parts are thus protected from corrosion both inside and out. This treatment is the major improvement to the Moke in 1988.

The bodies, before going in to the paint bays look quite attractive in their 'gold iridescent' finish! In addition to the zinc protection, box sections are wax treated and a chip resistant elastic undercoat is applied to both the inside and outside of the floor area.

The Moke progresses into the primer and final spray paint bays prior to entering the ensuing fitment bays. Suspension units, cooling system, engine and electrical equipments are added as the Moke passes down the assembly floor. At this stage it is possible to see the final destination of particular units - New Caledonia, France, Spain, Italy etc. Eventually the Moke, in pristeen condition, rolls towards the final trimming stage for the addition of seats and and a hood. Upon completion of this comes first start up and engine tuning. The vehicle is then complete and subsequently prepared for shipment to the dealer.

From the accompanying photographs some of these processes may be seen, together with a transporter loaded with Mokes. Jim is also seen with the latest Moke development of which more in a moment. The eagle-eyed amongst you will note that those vehicles destined for Italy have smaller, more widely spaced headlamps. This is to comply with Italian legislation and is yet another factor to be considered when catering for the international market.

Jim gave me a splendid insight into Moke production and updated me on some of the innovations in the pipeline. The anti-corrosion measures already described will be constantly tested by exposing a test vehicle to the severest of elements, notably salt water and salt spray. A new "Sunmaster" hood is on test and fitted to Salome's Moke. The hood is manufactured from an opaque plastic which has the properties of allowing passengers a sun tan without becoming sunburnt. The tensioning and securing of the hood has been updated and is definitely more efficient than the previous arrangement. Split rear seats which fold forward to allow a load space are now standard. This is not true for Mokes destined for France where the type is only homologated for two seats. Metallic paint is now an optional extra. Mono-colours are cheaper and of economic necessity, but the range available is extensive and of vibrant tones. Bumpers and grilles are now painted in 'White Diamond' as a real contrast to the body colour. The white on white version, that is a white body, hood, seats and wheel trims is most attractive. These innovations and developments enhance the concept of this superb vehicle - a fun car to own and drive.

Moke will probably take part in the "30th anniversary of the Mini" event to be held at Silverstone in August 1989. This will also mark the 25th Anniversary of the Moke. Jim is planning a limited edition Moke to mark the occasion and only 250 (10 for each year of production) will be produced.

This is an event not to be missed and will be a fitting tribute to all those who have ensured the success and continuation of this remarkable machine in the face of worldwide competition from imitators.

Cagiva Production Mokes

Jim Lambert's enthusiasm for the Mini Moke ensured it's survival beyond it's Austin Rover life. The manufacturing rights were put up for sale and he helped negotiate a deal with the Italian company Cagiva, a name famous for motorbikes, the Cagiva group including the famous marques of Ducati and Moto Morini. Jim then took a back seat as production of the Cagiva Moke began in 1991.

The model was essentially the same as the final Austin Rover Mokes - it was even made at the same factory in Portugal. Refinements were made to the Seats and Hoods and the vehicle featured (for the first time on a production Mini), a radiator mounted in front of the engine, just behind the grille. Other options available in the U.K. included new rounded bull bars, Minilite wheels, metallic paint and tinted windscreens.

Over production in 1992 led to the halting of manufacture in early 1993 and the closure of the factory in Portugal with the parts and tools being transferred to Italy. As always, rumours surrounded the future of the Moke: a new Italian factory and a 1.3 litre inject engine were two of the front runners - only time will tell if they proved correct.

Italians revive the Mini-based Moke

THE MOKE, ROVER'S Mini-based fun car, is to re-enter production — in Italy.

Manufacturer of the extrovert tin-top ceased in July last year at Rover's assembly plant in Portugal, where the Moke had been made since mid-1983 following its transfer from Australia.

Now Rover has sold manufacturing, sales and marketing rights to Italian engineering group Cagiva — makers of Moto Morini, Ducati and Husqvarna motorbikes and VM diesel engines, and the largest European manufacturer of motorbikes.

Production will restart next year, with a planned annual output of up to 2200 cars — roughly the same number built each year in Portugal.

Rover will continue to supply mechanical components for the Moke, which is expected to retain its 'A' Series engine until at least the end of 1992. The Italian car will be broadly unchanged, with the exception of a new front-mounted radiator in place of the Mini-style side radiator. This is to make the Moke comply with EC noise regulations.

Mini Moke production, suspended last year, is scheduled to re-start in Italy in 1991; UK imports are planned. More details as we get them . . .

The man in charge of the project will be former managing director of Rover Portugal, Jim Lambert.

"I feel very happy about the move to Italy," Lambert told *Autocar & Motor*. "Cagiva is very enthusiastic about the project and I feel the car now has a bright future."

Auto Express Poster
MINI MOKE

MINI we all know, but where on earth did Moke come from? It's actually slang for donkey, and no nickname could fit the bill better, because on its debut at the end of the Fifties, the Moke really did make an ass of itself, *writes Andy Wilman.*

Today it is synonymous with happy, smiley, groovy things like sunny days, peace and love, the Sixties, The Prisoner and cruising around Barbados beaches. But the Moke was never intended to enjoy this glamorous life of leisure.

Instead it was originally developed by Sir Alec Issigonis as a cheap utilitarian runabout for the army. In 1959 Issigonis handed over half a dozen prototypes, built on the chassis of his soon to be legendary Mini, for evaluation with HM Forces. The Moke was certainly tough enough, but wasn't used because of poor ground clearance.

It then served with the Navy as it was ideal on massive aircraft carrier decks.

Expensive

UK production lasted from 1964 until 1968, when government pen-pushers decided the Moke could no longer be sold as a commercial vehicle.

This meant the Moke was now subject to Purchase Tax adding an extra £78 to the £335 price tag.

It was now too expensive for a vehicle with no doors, and since 90 per cent of the 14,000 built had been sold abroad, British Motor Corporation transferred production to Australia.

The Aussies built it until 1982 and made improvements, such as toughening the suspension and fitting 13 inch wheels for extra clearance.

In 1982, production was moved to Portugal, where the Moke is still built today.

Now however, the Moke is being imported again by official UK concessionaires Duncan Hamilton of Hampshire, and also Fullbore in London. The 1993 version has better brakes and handling, plus low running costs.

It has a 1.0 Mini engine, an better hood and lockable luggage compartment.

Sales continue strongly in hot countries, and this year the Moke will get Rover's fuel injected Cooper S engine.

● Contacts: *Duncan Hamilton 0256 765000; Fullbore 071-371 5931.*

FEELING GROOVY: The Moke evokes images of summer days and the Sixties

DATI TECNICI

MOTORE
- Anteriore trasversale, 4 cilindri in linea di 998 cc.
- Alesaggio x corsa: 64,58 x 76,20 mm.
- Rapporto di compressione 8,3:1

TRASMISSIONE
- Trazione anteriore
- Cambio meccanico a 4 marce sincronizzate

SOSPENSIONI
- Indipendenti
- Elementi elastici in gomma e ammortizzatori idraulici telescopici

FRENI
- Sistema idraulico a doppio circuito
- Anteriore a disco, posteriori a tamburo

RUOTE
- Cerchi in acciaio 4,5 B x 12
- Pneumatici radiali 145/70 SR

INFORMAZIONI GENERALI

- Altezza 1.460 mm.
- Lunghezza 3.232 mm.
- Larghezza 1.440 mm.
- Passo 2.038 mm.
- Altezza da terra 190 mm.
- Diametro di sterzata 9.400 mm.
- Peso 630 Kg.
- Portata 400 Kg.
- Capacità serbatoio 39 l.

PRESTAZIONI

- Potenza massima: 29 Kw a 4759 giri/min.
- Coppia massima: 68 Nm a 2500 giri/min.
- Velocità massima: 112 Km/h
- Consumo: litri 7,5 x 100 Km (90 Km/h)
  litri 7,0 x 100 Km (urbano)

La Moke si riserva di apportare modifiche senza preavviso.

TECHNICAL DATA

ENGINE
- Transverse, 4 cylinders, 998 cc.
- Bore x stroke 64,58 x 76,20 mm.
- Compression ratio 8,3:1

TRANSMISSION
- Front-wheel drive
- 4 Speed manual gearbox

SUSPENSION
- All indipendent
- Rubber cone springs, telescopic shock absorber

BRAKES
- Dual-line hydraulic system
- Disc front, drums rear

WHEELS
- Pressed steel 4,5 B x 12 wheels
- 145/70 SR low profile radial ply tyres

GENERAL INFORMATION

- Height 1.460 mm.
- Length 3.232 mm.
- Width 1.440 mm.
- Wheelbase 2.038 mm.
- Ground Clearance 190 mm.
- Turning Circle 9.400 mm.
- Weight 630 Kg.
- Max Towing Weight 400 Kg.
- Fuel Tank Capacity 39 l.

PERFORMANCE

- Max gross power: 29 Kw at 4759 rpm
- Max Tarque: 68 Nm at 2500 rpm
- Top speed: 112 Km/h
- Economy: litres 7,5 x 100 Km (90 Km/h)
  litres 7,0 x 100 Km (urban)

Moke reserves all rights to apply modification without notice.

MOKE AUTOMOBILI S.P.A.

Commercial Office:
Via Caronaccio, 67 - 21040 MORAZZONE
VARESE (ITALY)
Tel. (0332) 462875 - Fax (0332) 462877 - Telex 380280

CAGIVA GROUP

Mokes Off Road

Considering its origins as a light weight jeep for the army, it is not surprising that the Moke finds itself equally at home away from the tarmac. The initial demonstrations by BMC included hurtling it around military test grounds up to its axles in mud and snow. And although the army decided against the Moke because of its poor ground clearance, Mini Moke Clubs and owners have continued to use it for its original purpose, and more. Some, such as the South Australia Moke Club take it to excess!

For more leisurely off road pursuits, the Mini Moke served a practical purpose for farmers and landowners, driving around fields carrying equipment, hay, and even the occasional sheep if required. They also became popular as golf caddies, before the electric car became more practicable, with golf clubs such as the Royal and Ancient at St Andrews in Scotland owning fleets.

Interest shown in the single- and twin-engined Mini Mokes by major oil companies and large overseas civil contractors prompted B.M.C. to demonstrate these vehicles recently at Bagshot Heath. Both the rough road and the 'alpine' circuit of the military test ground there were utilized, and the 'alpine' was made particularly spectacular as a result of rain. Two four-wheel-drive Gipsy vehicles were also exhibited and carried representatives over the courses, whilst members of the design team and the Export Corporations of Austin and Nuffield were present to answer marketing questions.

A single-engined, 850-c.c. Mini Moke, despite only having two-wheel drive, proves its versatility under extremely difficult conditions on the rain-soaked 'alpine' track

(Bottom left) Austin Gipsy and Mini Moke tackling the Bagshot Heath vehicle 'assault course' in fine style

(Bottom right) The two-engine Moke startled several passengers with its 'Magic-Mini' performance, frequently showing the ubiquitous Austin Gipsy models the way round the slippery course

Unsuitable for motors

DOUGLAS ARMSTRONG takes a Mini-Moke along the Ridgeway—a road that was built long before the Romans came to Britain—where King Alfred defeated the Danes

THE Ancient Britons 'built' the Ridgeway along the top of the Berkshire Downs much more than a couple of thousand years ago, and in spite of the surface they would have undoubtedly welcomed Mini-Mokes for transport. Nowadays, the road—if you will pardon the expression—is much the same as it was in those halcyon times, and although the various access roads are nearly all marked 'Unsuitable for Motors', I knew the Mini-Moke would treat the statement with the contempt it deserves.

The Ridgeway winds along the top of the Downs, presenting fabulous views, eventually joining up with the ancient Icknield Way which, as the Upper Icknield Way, becomes Wendover's High Street (B4009/B4010) in Bucks. The Icknield Way, of course, has got itself 'metalled' over the years, but the Ridgeway is still the grassy, muddy, rocky, chalky track that it ever was, and if you ever feel like getting away from it all it is worth tracking down. Get into the Thames Valley between Streatley and Blewbury, and if you are driving toward the latter Berkshire town it is up high 'on the left of you'. Parts of it are smooth and grassy, but some are rutted and tricky—Mini-Moke country in fact. When I introduced 'my' Mini-Moke to this ancient terrain it just could not wait to get on with it. It tore over the grass, rushed through the mud, and treated the ruts, potholes, and ridges with twentieth-century disdain. The Moke was fitted with perfectly standard tyres and yet it seemed able to go anywhere. With the optional tracked tyres on the driven front wheels this agricultural variation on the Issigonis theme would probably motor up the side of a Roman fort. After a morning of mud and ridge-storming the Moke idled as though we had just been down to the Post Office to spend fourpence. It was not hot or distressed—but it was certainly dirty.

The Moke's engine/transmission unit is, of course, identical to the Mini-Minor's—same cubic capacity (848 c.c.), same compression ratio, same power output, everything. It is even geared the same at 14·824 m.p.h. per 1,000 r.p.m. in top gear. With only 18 inches of the total length of the vehicle taken up by the engine and transmission there is room for four men and impedimenta. Of course, its 'brick wall' frontal aspect can hardly be described as streamlined—there is no reason why it should be—and the top speed seemed to be around 60 m.p.h. A happy cruising speed on the road was an indicated 50, at which velocity the little engine would purr away and the car would take the corners in true Mini style.

Practical features

The car bristles with practical features: like a bonnet which not only hinges upward for engine inspection but lifts off with a pull at the rubber clip/hinges; massive

steel grab handles for all passengers (very comforting on cross-country journeys!); quickly detachable panels in the boxed main structure for inspection of battery, fuel pump, and so on; and removable seat cushions for cleaning and drying. The large hood could be put up or down in a flash and provided a reasonable measure of protection when it was raining, but for winter use I think I would like a set of sidescreens (these are available from accessory manufacturers) and a heater.

I took a friend with me on Operation Ridgeway and he was as impressed as I was with the surefootedness of this cheeky little vehicle. Some of the time the Ridgeway's ridges were so deep that the sump was skating along the muddy surface. The optional extra sump guard is a worth-while investment, if not a necessity if you are to use your Moke off the beaten track.

Our off-the-beaten-track run took us on to the Icknield Way, across the Downs to Wantage, that pleasant town where King Alfred the Great was born 1,116 years ago, and where, presumably, he burnt those legendary cakes. The cake-burning of the great Saxon may or may not have happened, but he certainly defeated the Danish hordes on the near-by Downs, whereupon they retreated to Reading which was a Viking stronghold at that time. If Alfred had had a few Mini-Mokes the course of English history may have taken a completely different turn.

Averaged 37 m.p.g.

Yes, Alfred would have been impressed even more than we were, but after rushing about all day on track and road we found the Moke had averaged around 37 m.p.g., and had used not a drop of oil. On the road, acceleration was of the order of 0–50 m.p.h. in 15 seconds two-up, and there is little doubt that all the figures will improve with the passage of time, for the car had a very low mileage on the odometer. Definitely the wear for farmers, builders, bathers, Ancient Britons, and Saxons.

After a bout of very wet weather the Moke would start first jab, even though it spent its nights in the great outdoors—just as it was designed to. The warm-up from cold was rapid due to a well-suited thermostat, and the engine soon settled down to a reliable idling speed. After a few days with the Moke I was genuinely sorry to see it go—not to mention Fred, my Labrador, who regarded it as very huntin', shootin', and fishin'.

Doubtless the Ridgeway will be there for a few more hundreds of years for keener types to try their cars (and motor-cycles) cross-country efficiency, but strangely enough it would seem that the latest extension of that ultra-modern road, the M4, will start to stretch its concrete spans across the ancient Berkshire Downs. Quite a contrast.

Mighty Mokes take on the Outback

"We'll drive them anywhere and bring 'em back alive," say the members of a club who have ranged far and wide in their "cheeky little chariots". By Greg Rinder, as told to Liz Johnswood. Pictures by SA Moke Club.

AUSTRALIA'S early explorer, Sturt, knew exactly what he was talking about when he said, "This land will not tolerate the weak."

This was the first thought that flashed into my head when I saw the green Moke flying towards me. The next wasn't so profound. I thought: Hell! I hope it's Phil's green Moke, not mine!

The year was 1970 and Phil Heenan and I were on our first trip into the outback from the South Australian Moke Club. We'd parked our Mokes at the top of a high sandy bank at Policeman's Waterhole on the Cooper Creek, and were down below clearing a spot to make camp.

We heard a rush of wind and a crashing noise and there was the Moke hurtling at us. One of those sudden willy-willy winds you get in the outback caught it and tossed it over the cliff like a kid's toy.

It was my Moke, as it happened. Near the bottom of the cliff, it hit a huge gumtree and the roll-bar stopped it from going further. If it hadn't been for that, it would have gone into a 13m deep waterhole.

It was a piece of good luck, I guess. That's if you can call it good luck to have your new Moke up a gumtree 1500km from home.

We'd been heading up the Strezlecki Track from Adelaide to Innamincka, the old border outpost on the Cooper, in the far north of SA. Our destination was close by, so we drove Phil's Moke in and borrowed a cross-cut saw.

I was sure my Moke was a write-off and I was glumly thinking, well, at least it met its end on famous ground — half way between the graves of Burke and Wills, the explorers who died so tragically not far away in the early 1860s.

"We're doomed! We're going to perish! We'll never get over the sand dunes and saltpans"

But you can't keep a good Moke down, we found. After sawing it away from the tree I tried the engine and it kicked over. We inched it up the bank and I went on with our trip sporting no roof, no windscreen, a bent steering wheel and drastically remodelled bonnet.

That first disaster should have cured Phil and me of poking our Mokes into places they're not designed to go.

"You need big, four-wheel drive vehicles, not two-wheel drive butter boxes like yours, to get through country like this," the outback experts kept reminding us.

The warnings and our first trip had the opposite effect. They whetted our appetite for battling our little Mokes through hardships and inspired a saying in our club, "We'll drive them anywhere and bring 'em back alive."

It's a wonder they brought **us** back alive, though. To start with, we went north in the middle of summer. It was around 43 degrees Celsius in the shade the day we left Adelaide, two greenhorns in two green Mokes in that wide brown country.

The most naive bushie knows you can literally fry an egg outside in the December sun and dehydrate as fast

Top: Bushman Jim Forrest enjoys a plate of rabbit stew, Moke-style. Above: Driving down a railway laid straight along the dried-out Finke River. Opposite page: A broken-down Moke. On one trip, all the Mokes went out of action.

as a fish out of water. They could have told us summer is cyclone time up north and when it rains it rains buckets — enough to cut you off from civilization for weeks or sweep you away in a sudden flash flood.

As well as going at the wrong time, we took the wrong things — like 20 tins of camp pie that poured out of the tins after the heat got to it. True to form, we left the right things at home — like plenty of spare parts for our Mokes. We had a spare fan belt, a radiator hose and that's about it.

The following year, on an even longer Moke-about to Alice Springs in the centre, we repeated the mistakes of the first year perfectly.

We left on Christmas day with our spare fan belt and

radiator hose. Our mechanical knowledge was as meagre as our sense. If you turned a key and the Moke started, it was OK. If it didn't, it was broken.

As the kilometres piled up in the heat and the dust our Mokes started making horrible noises. We ignored them. I had a shot wheel-bearing and didn't realise it. At one stage the temperature in my Moke went up to 54 degrees Celsius and off the end of the gauge.

What insane innocents! We didn't deserve to make it to The Alice but we did, the last 1200km over rough dirt roads which were par for the course in the early 1970s.

On the run back I had to stop every 15 kilometres or so and wait 20 minutes for my Moke to cool down. Oddly enough, the wheel didn't fall off.

When we limped into Coober Pedy, along with a dust storm, a chap came up and said, "You'd better keep moving or you won't get through. Big rains are on the way."

Crazy bushie, we thought. It hadn't rained in Coober Pedy for seven years and we could see no signs now. The long drought had obviously got to the fellow. We were so dog-tired, hot and dirty, we didn't want to know about anything other than a shower and a sleep.

Next morning when we woke, everything was under water. Cars were bogged in the main street. Roads were cut in all directions. The locals were happily paddling around in gumboots.

By the time we sloshed and mud-slid our Mokes home to Adelaide, the innocents weren't so innocent any more; and we had a new respect for our bushie and this wonderful, unpredictable country of ours.

People tell us they can't understand why we go Moking outback. Punishing ourselves, as they call it. It's no mystery to us. Ever since we began our Moke club in Adelaide in 1969, we've been fascinated with Australian history.

Our trips are planned around the routes taken by early explorers and pioneers. We research their lives and run off mini-biographies for each Moker to study on treks.

We're aware we're getting it easy compared to those incredible men but there's something about a Moke that keeps you in touch. The ground's close, running along beside you, and a breeze blows over you, bringing the taste and smell of the bush.

It really gives you the feel of a place and the people who have gone before. At Innamincka Station I could almost see the Cattle King, Sidney Kidman — a thin 13-year-old setting out to make his fortune with five shillings in his pocket and riding his old one-eyed horse, Cyclops.

Admittedly, we're all fascinated with Mokes, too. Someone wrote: "Moke addicts fall into two categories; owners who can afford to run a Jaguar but don't and owners who can't afford a Jag and don't care. And both have a continuing love affair with the cheeky little chariots." They're right. There's a mystique about Mokes that gets to you.

By 1973, thoroughly familiar with our "cheeky little chariots" and with two chastening trips under our belt, we considered ourselves the boy scouts of moking — bush wise and prepared.

With almost enough spares to rebuild a vehicle we wheeled off in perfect September weather and did a 2800km, trouble-free run up to famous Birdsville on the edge of the Simpson Desert and back.

A character from that trip stands out — a man who'd obviously been propping up the bar in Birdsville for a fair period. He came burning up in a broken down Holden station wagon, stopped, staggered out and thumped a Moke on the bonnet. "First covered wagons I've seen for decades," he told us.

In April, 1974, jacked up with last year's success, we thought we'd do an easy run up to Andamooka, one of SA's famous opal mining towns, 600km north of Adelaide. We were promptly knocked off our jack with the notorious floods of that year.

At one stage our Mokes became separated and I found myself in the dark, stranded in the middle of a lake that had never been there before. We learnt later it had already claimed the motorbike of the local policeman on his way to Maree to visit his girlfriend.

My offsider was Graham Shepherd, a chap we called Hardluck. He'd never been bush before and he sat up all night with a shotgun, scared stiff.

We waded to a piece of higher ground and sat ringed with dingoes, their eyes gleaming in the light of a torch as we swung it around. Hardluck kept saying, "We're doomed! We're going to perish! We'll never get over the sand dunes and saltpans. We'll run out of water." We were surrounded by 50 million gallons of the stuff.

Around two o'clock in the morning one of the other Mokers, Keith Norgrove, came staggering up with a torch and two bush biscuits. In line with our club rule

"We drove into the terrible country of Sturt's Stony Desert ... it was a bone and soul shattering experience"

of keeping track of one another, he's walked for hours to find us. By the time we got back to Adelaide another lesson had been well and truly hammered home — the best laid plans of Mokes and men are no match for the Aussie Outback.

Mokitis can strike anyone, anytime, so the mokers on our trips are a very mixed bunch — tuna fishermen, micro-surgeons, bricklayers, hairdressers, nurses, entomologists, soldiers, policemen, and truckies.

Age is no protection from the disease as we found out in 1976 when we tackled an ambitious trek to Ayers Rock and Liz Rundle brought her mother, Lady Jude (wife of Sir Norman Jude) on the trip.

To get into the trek, the good lady had whipped 10 years off her age and said she was in her fifties. She turned out to be well into her sixties and we wondered what we'd let ourselves in for.

We needn't have worried. Lady Jude was more than a match for most of the blokes. She helped pull Mokes through the Neales River, 40m across, a metre deep and flowing, she tackled 700km of flooded outback roads and sat around a campfire at 3am cheerfully drying out after a mini-cyclone had flattened and swamped 11 out of our 12 tents.

She was always first up in the mornings getting the campfire going; but the classics were when she polished her little red Moke until it shone after each muddy day's run and opened a well-stocked "bar" before dinner each night and invited everyone for drinks. She taught us all a lesson in surviving in the Outback in style.

The characters you meet out there are equally as large-as-life. We still talk about an oldtimer we met in a pub in Birdsville. One of our chaps, Warwick Williams, got yarning to him and said, "Are you a shearer?"

The old fellow held up his hands, disgusted. "Do these look like the hands of a shearer?"

Not to be outdone, Warwick held his hands up. "Do these look like the hands of an oil refinery worker?"

"Whatever you do mate, you don't do much of it," the oldtimer came back smartly.

I remember during a trip out the back of Bourke in NSW, a dozen of our Mokes came cruising over a hill and almost ran into a bloke sitting staring at the shattered trailing arm-pin of his Mini-Moke.

He just gaped at us. I'm sure he thought a dozen Mokes way out there were a mirage. We whipped a trailing arm-pin from spares, fitted it on, handed him a cold beer and were on our way before he came out of shock. He's probably still telling his mates about his hallucination in the desert.

Another character we've enjoyed over the moking years is Mike Steel, who runs the trading post at Innamincka.

Mike, known as The King of Cooper Country, has made a great contribution to the remote area by signposting all the Burke and Wills' monuments. They wouldn't be accessible for history buffs like us if it weren't for men like Mike.

We called on him again in 1977 while following the route taken by Sturt on his 1844 expedition to find Australia's Inland Sea. Drinking in the cool of Mike's little store, we wondered what Sturt would have given for a Mike Steel to turn up in that merciless territory.

When we left Innamincka for Birdsville and drove into the terrible country of Sturt's Stony Desert, it was a bone and soul shattering experience. How Sturt felt can barely be imagined when he found his Inland Sea sea of sand — the Simpson Desert.

Our 3500km trip took 10 days. Sturt's took two heartbreaking years — and he left his friend and second in command, James Poole, in a lonely grave along the way.

We tasted failure ourselves the following year when we followed the route of another legend — John Stuart, whose expedition across Australia paved the way for the overland telegraph.

We drove up from Adelaide to the western border the Simpson in an attempt to reach Chambers Pillar,

Above: Hauling a Moke across the Neales River on the Ayers Rock trip. Left: More water trouble. A crocodile took a horse near this spot on the Wenlock River, Qld.

the historic landmark for explorers who have left their names blazed into the face of the rock.

After days of bulldust, corrugations and gibber stones, dragging Mokes across a 300m stretch of soft sand (the dried out Finke River), getting bushed because our Army Survey maps were stolen somewhere along the track and running dangerously short of water, we had to throw it in just 8km from our target.

We tackled the Pillar again later that year and made it. When we finally came on the weird towering white column capped by red sandstone, it was worth every body and nerve-stretching metre of the way. It looked like a conqueror standing there in the crimson desert.

I've been truly spooked only once in our moking around Australia — crossing the Gibson Desert in 1980 along the famous Gunbarrel Highway.

Out in the sand dunes in 50 Celsius heat, that poem of Barcroft Boake kept running through my mind;
Out on the wastes of the never never,
That's where the dead men lie,
There where the heat waves dance forever —
That's where the dead men lie.

Gibson wandered into this desert in search of horses during the Giles expedition. He was never seen again.

Above: Going for broke in a Moke across rough, outback country. **Left:** After this roll-over, the Moke was flipped back on its wheels and driven. Mokes are tough!

The thought that we may be driving over the place he died, really got to me.

It was brought home to us just how vulnerable you are in territory like this. It was an effort to move and we had to drape wet cloths around out heads to get some relief. Our battery-run fridges packed up. Drinks came out at body temperature.

Everyone was showing signs of dehydration and our first-aid man was busy handing out extra salt tablets. One of our fellows, Paul Huxford, hallucinated and threatened to kick his windscreen out.

We were lucky to have an experienced bushman, Jim Forrest, with us. He brought Paul back to normal with the old bushman's trick of pouring a bucket of our precious water over his head.

We thanked our lucky stars for strict club rules on that trip. It's compulsory to carry around five litres of water per person per day, food rations to last a whole trip plus extra for hold-ups.

Boots instead of thongs are worn so snakes can't get a free go at your toes. The big King Browns and aggressive Taipans are around in some of the country we cover as well as the "fierce snake" (as it's called by locals), said to be the most deadly in the world.

Our Mokes are modified to club specifications, we

"It was an effort to move and we had to drape wet cloths around our heads to get some relief"

carry CB radio, full first-aid kits and recommend before a trip that members join St John Ambulance, which covers them for flying doctor services throughout Australia. If you're not covered you could be up for fares up to $3000 just flying out from remote areas.

"A croc took a horse here last week," a local informed us after we'd man-hauled our Mokes across the Wenlock River on the way to Cape York, at the top of Oz, in 1982.

A bunch of four-wheel drive characters ran a book on whether we'd make it to The Top — 100 to one against.

Maybe they were omens that this marathon was going to be our Everest — our most fascinating yet devastating trip yet. One moker, Barry Smith, became sick and had to be flown out and, at one stage, all four Mokes in the team were out of action.

One lost a wheel on the way up and had to be piggy-backed by transport to civilisation for repairs. On the home run another ran backwards over a cliff after the brakes failed.

Again a friendly gumtree came to the rescue and stopped the Moke from plunging beyond recall with Judy Loudon and Neil Spouse on board. The intrepid pair crawled out unscathed. The two remaining little battlers broke their axles trying to tow their friends to Cairns. And so there were none.

Yet the trip was a triumph, in spite of the dramatics. We reached The Top and after repairs to our Mokes, made it home to Adelaide — 10,000km in 23 days, 11 of which were lost in delays and breakdowns.

Our grand trek of 1983 — driving nine Mokes with 18 people across the largest sand-ridge desert in the world, the mighty Simpson — was a dream we'd been building ever since those greenhorn days.

Every moker in the team admitted to some qualms. That super-Simpson expert, Dennis Bartell, and map fanatic, John Whitburn, put some backbone into us and made the whole thing feasible.

From Purni Bore to the famous Birdsville Track, across a thousand sandhills on tracks the desert is slowly reclaiming, after three days and 700km of concentrated driving, we got there. There's a commemoration plaque to prove it behind the bar of the Birdsville pub.

Where do we go from here? Some moker, someday, will spot a place on a map and get a yen to see it. A pioneer's imagination will be fired and we'll be off.

Canning's Stock Route in the Northern Territory, where it's 1600km between petrol pumps, is one yen. Just carrying fuel for that trip sounds impossible. But that's what our Moke Club's all about. ●

Mokes at historic Chambers Pillar on the Simpson Desert. Explorers left their names blazed into the rock's face.

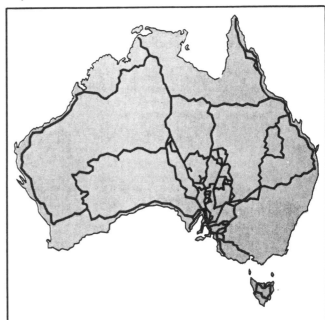

The map shows routes taken by the South Australian Moke Club on their trips. They began in 1970 and, on their last trip, crossed the daunting Simpson Desert.

THE WORLD'S MOST REMOTE OFF-ROADER

Pitcairn looms over the schooner's bow, 3300 miles east of New Zealand, 4100 miles west of South America. It became the setting for the world's most improbable off-road test.

PHOTOS BY SPENCE MURRAY AND WALTER BERSINGER

At something less than full-chat, the Moke exits Turn No. 1 with, if not alacrity, student power in reserve.

Unwashed except by rain, unloved and largely unused, the currently immobile Moke reflects the hard life on remote Pitcairn Island.

By Spence Murray

DOES ANY VEHICLE become an off-roader if it's used only where there are no roads? Your worthy editor, Duane Elliot, and I were discussing this one day over lunch, and it was decided to let the readers be the judge after perusing the following tale.

It so happens that in the mid-Pacific Ocean, halfway between New Zealand and South America, and 1500 miles south of the equator, lies a lonely, tiny speck of land. This is Pitcairn Island. And if you've read of the famous mutiny aboard the ship *Bounty* 200 years ago, or seen any of the five major motion pictures based on the subject, then you know about Pitcairn having been the destination of the mutineers under Fletcher Christian after they had cast the infamous William Bligh and his followers adrift in mid-ocean.

To make a long story mercifully short, understand that Pitcairn today is inhabited by fifth and sixth-generation descendants of the *Bounty*'s miscreants.

Pitcairn is and always has been the most isolated, least-visited and most difficult place to reach on our planet. And the *Guiness Book of World Records* notes also that it is the world's smallest colony. The island is unique in that it has neither harbor nor airstrip, and comes closest to being the place you can't get to from here. It's a tortuously steep volcanic mass of only 1120 acres, and rises 1000 feet above the sea with 27 percent of its area unscalable cliffs.

What does this have to do with off-roading? Just that it took me two years of logistical planning and fighting bureaucratic red tape to get ashore on the lonely island. After the time, trouble and expense devoted to such an unlikely road test, *Off-Road* would take any car that Pitcairn had to offer. If it had a car.

Your unworthy scribe reached Pitcairn, as noted, only after much planning and confused negotiations. One must have written permission from Pitcairn to stay more than a few hours, and the mail being what it is, this can consume a year. Then you have to have approval from New Zealand, which has jurisdiction over the island. Finally, one must deal with the French, agreeable-enough folks on their own side of the globe, but who take exception to foreigners passing through their nuclear testing zone, as I would have to do aboard a cramped schooner out of Tahiti, 1250 miles from Pitcairn.

The wheel came early to Pitcairn with the advent of a peculiar barrow designed to cope with the island's steep, red-clay pathways. The trails are used for the carting of fruit, produce from scattered gardens, and timber for house-building. So precipitous are the hillsides that a luckless cow sent there during the 19th century fell 350 feet into the sea. Ages ago, a few scrawny horses were sent over from Mangareva, the nearest inhabited atoll, some 350 miles distant, but they died from an undiagnosed illness, which is why they were relocated in the first place.

So Pitcairn has never enjoyed beasts of burden, and thus, had no use for vehicles for them to pull. Commerce was carried out afoot, and with rough paths totalling only 3.6 miles on the island's 1½x2-mile area, walking was the easiest way of getting around. It has always been easier to plod, scramble up and slither down the narrow trails than to move about by any other means, and it wasn't until the 1960s that the first motorized dirt bike arrived to usher in the advent of the horseless carriage.

At the dawn of the 1970s, a well-meaning island administrator in far-off New Zealand suggested that Pitcairn should have a proper car to enable the elderly to get around. The idea was met with excitement, and a Toyota Land Cruiser was picked from names dropped in a palm-leaf hat. But cost-conscious overseers 3300 miles away prevailed, and it wasn't until a year later that a ship came over the horizon bringing Pitcairn's first car. It was, of all things, a Leyland-built Australian Mini-Moke.

The logic behind this, cost aside, was the island's inability to unload from an offshore freighter anything beyond the one-ton capacity of the crane at the jetty in Bounty Bay. But it seemed reasonable to assume the almost-car's front-wheel drive and golfcart-sized tires would enable it to skeedadle up Pitcairn's inclines with some degree of alacrity, and with mechanicals straightforward enough to fall within the scope of the island engineer.

Wrong. Rainfall averages 80 inches annually and turns the rutted paths

THE WORLD'S MOST REMOTE OFF-ROADER

People-power supplants horsepower as Education Officer Lloyd Buckley, left, lectures some of his Pitcairn students on the care and feeding required by an automobile.

The author at the helm of the world's most remote car. Only the Lunar Rover left on the moon is further from civilization—and just as inoperable.

into sticky quagmires, so the Moke could only spin its front driving wheels hopelessly. The machine was parked in a grove of banana trees where it was later consumed by fire. Undaunted, Administration sent out a second Moke, and while it was useful during one dry season (July and August), it was ultimately abandoned to terminal rust, though some of its innards are occasionally cannibalized to keep Pitcairn's third automobile up on all fours. Folks, meet the island's third Moke, and without question the world's most remote and loneliest off-roader.

I reached Pitcairn (thus relieving the readership of the harrowing 1300-mile trip by aging schooner with French gunboats trailing us on occasion), at the only landing place of Bounty Bay. First thing I did was inquire after the ubiquitous Moke. The jolly Pitcairners (who speak a softly-slurred 18th century English, by the way) seemed puzzled until someone recalled there was a car somewhere on the island. If he could only remember where.

At length, the Moke turned up in a shed near the schoolmaster's house at the base of a towering cliff which harbors, half-way up, the cave which tradition holds was used by Fletcher Christian to meditate the errors of his mutinous ways. The nearby school has a present enrollment of 13 (out of a total island population of 47), in answer to British law requiring formal education of all children. The present school teacher, on a two-year rotation out of New Zealand, oversees Crown property on Pitcairn, and it was in his shed that the Moke resided.

The unloved car, it turned out, didn't run. Something about a missing battery, fouled carburetor, dubious brakes and assorted other ailments. To make matters worse, the Moke's shed was atop a steep but (luckily) paved path (the only real driveway within 1300 miles). The vehicle was in service as a school storage bin, and not only that, but intermittent showers were being driven by 35-knot winds. But quite undaunted, and anxious to have the Moke under his care featured in a world-renowned magazine (of which he had never heard), teacher Lloyd Buckley called a school holiday. Eager hands exhumed the Moke from what all had hoped would be its final resting place, to sit poised on the brink of the slope.

Pitcairn's only eighth-grader was assigned to pilot the Moke during its dizzy descent to a patch of level ground after a hasty lesson in braking and steering (due to a natural distrust of an alien—me—with Empire property). Should the former fail, then the latter might be useful for avoiding

Jetty at the landing place provides little protection from the surf, which, at times, crashes heavily against the shore.

Of little use is the transverse 61.0-inch four-cylinder engine, its 40 horsepower (while running) scarcely able to cope with the island grades, and its front-wheel drive a poor choice for wet clay.

coconut trees and the schoolhouse, and to veer the Moke harmlessly into a dense thicket, rather than have it plunge into the sea. With emergency braking provided by the rest of the student body, and amid the cheers of a growing rooting section, the hapless car was somehow safely deposited at the foot of the slope where it wouldn't roll away of its own accord.

As the representative of *Off-Road*, it was now my editorial duty to make the most of a desperate situation, wind and rain nothwithstanding. Two year's preparation and seven weeks, involving 8200 air-miles and another 2700 by schooner, would certainly not go for naught. So while photographic apologies are in order, I make no excuse for the fact that my test drive was limited to back-and-forth runs of 15 feet, while the students, their teacher, and the police officer pushed me to-and-fro until, consumed by laughter, they could push no more.

Bottom-line data on the curious Moke is detailed in the accompanying chart, yielding the secrets of a standard Moke alongside those of Pitcairn's example. And so having thus fulfilled my promise to the good people of Pitcairn to popularize their fantastic little island among the world's motorists, I must relegate a copy of this edition of *Off-Road* to the long and problem-ridden trek to Pitcairn where, with luck, it may arrive in as little as six months' time. Bon Voyage! **OR**

PITCAIRN MOKE SPECIFICATIONS

GENERAL	THEORY	ACTUAL
List price	n/a	Free
Curb weight	n/a	Too heavy to push
Wheelbase	83.5 in.	Too long
Track, front/rear	49.0/49.75 in.	Raining, couldn't measure
Length	127.25 in.	Hardly matters
Width	57.0 in.	Too wide for path
Ground clearance	8.0 in.	Insufficient
Fuel capacity	6.0 gal.	Tank removed for other use
ENGINE		
Type	OHV inline 4	Who cares?
Bore/stroke	63.0x68.3mm	Raining too hard to check
Displacement	998cc	Can't prove it by us
Horsepower	40 bhp	13, when school's out
Torque/rpm	50 lb-ft	Unsustained
Induction	semi-downdraft	None
CHASSIS AND BODY		
Layout	Front engine/front drive	Not this one
Body/frame	Integral	Makes good chicken coop
Brake system, f/r	2 front leading shoes, lining area 74.0 in.	Shoes and bare feet
Steering	Rack & pinion	Luckily, it works
Suspension	4-wheel independent, rubber springs, telescoping shocks	Simply awful

Moke Club members tackling the elements.

Racing Mokes

With a body shape described as a "bedstead on wheels" and the aerodynamics of a "brick", the Moke hardly seems a likely racing vehicle, and indeed the English Moke never succeeded in reaching 70 m.p.h. in the original road tests. However, the Moke will lend itself to anything and the addition of a suitably sized engine and the removal of the windscreen improved matters dramatically.

The Moke was most at home in Autocross, and the BBC Autopoint program provided the perfect backdrop for it to show off its talents. This was an annual competition held between the Army and the London Motor Club and was broadcast on the Grandstand program (see pages 72 to 75 for pictures). In addition John Player & Sons had four special Mokes prepared (The Players No. 6 1275 Mini Mokes) which gave a surprising performance when thrown around a mud course.

Further afield, as the following articles show, the Moke has been used on international rallies, and raced around circuits in various parts of the world.

Supercharged Hill Climbing Moke ('Spoke") - HOA 124D

This was a very special racing Moke. In the late 1960's this Moke made many appearances at Sprints and Hill Climbs (Prestcott, Shelsey etc.), raced by Reg Phillips and former BMC publicity manager Raymond Baxter. They report that it gave them very little trouble and an enormous amount of fun. The engine itself was rather special. A supercharged 1275 'S' with polished crank and rods etc., it developed 150 horse power. The gearbox was straight cut with a very low limited slip differential to stop the torque from the engine destroying the gearbox. The supercharger was a Sir Godfrey unit which was normally used as a cabin pressuriser for a Comet airliner.

The vehicle is currently under restoration and will be returning to the racing scene later in 1989. The photograph below was taken at the start of the Shelsey Climb in 1968. The driver is Reg Phillips.

MUDPLUG '68

The annual 'battle' between the Army and the London Motor Club was fought recently over the cold, windswept wastes of Hungry Hill—a desolate, tank-testing course within a Sergeant Major's shout of Aldershot. The brave mortals (and machines!) who compete in the BBC Autopoint mudplug are seen by millions of Saturday afternoon Grandstand viewers, who watch the cars and service vehicles in action from the comfort of their armchairs. The Army—represented as always by the British Army Motoring Association—won by a mere four points—bringing the annual tally to four-all. The Mudplug is a team contest, the two seven-a-side teams having the alternative choice of starting positions—the object, of course, to be first across the finishing line. Between plunging from the starting plateau and storming up the final hill, men and machine go through an ordeal of flying over humps and bumps, dashing through mud bowls of various depths and speeding along the sandy stretches at

speeds over 60 m.p.h. This year's contest marked the first appearance of the Ant—a four-wheel-drive, 1275-c.c. transverse-engined Austin projectile, which notched up two victories and even finished second in another race after being pushed back on all four wheels by the exuberant driver who went too fast up a steep adverse bank! Famous rallyman David Seigle-Morris flung the versatile Mini-Moke round to good effect, but did reach the altitudes of Staff-Sergeant Mick Soames in the Army Moke, who helped the Army to seal victory in the last race of the day. Well-known autocross exponent COLIN TAYLOR, who took all the excellent photographs which appear on these pages and overleaf, competed in both an Austin 1800 and a Wolseley 18/85, and told us later that both cars felt completely at home in the rough conditions.

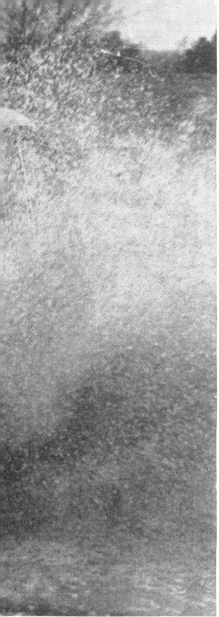

Not nearly as fast as it looks. Cardno getting crossed-up.

No Moke without mire

A very private autocross

by Hamish Cardno with pictures by Maurice Rowe

WE'D spent a very wet and muddy Sunday at the Player's No. 6 Autocross Championships at High Wycombe, 'way back in September, when one of the side attractions was the Player's No. 6 Autocross Mini Mokes being driven by "invited personalities and journalists". This nice man said: "Would you like to have a bash at one of the Mokes?" and I said "Yes", wondering all the while if it really was such a good idea, as girl friend couldn't drive and I had to get back to the office that night with 2,000 words plus results. And if I didn't get back there'd be two blank pages in *Motor* and a sudden vacancy on the staff.

Things went wrong though, because it got wetter and muddier and the actual autocross finals got a bit delayed, and the chief marshal said: "If you don't mind lads, drive these things some other day—the marshals want to get home" (or words to that effect). So we didn't.

Then about Tuesday the nice man phoned up and said had we got the mud wiped off us, and I said yes, and would we like to come down to a field in Surrey on Friday with "a few other invited journalists" and drive the Mokes. Again I said yes.

But here comes the catch. Until the Thursday morning I had thought this would be a great caper. Cardno versus a couple of clots from the *Budleigh Salterton Clarion* who only drive about five miles a week, and that in some dilapidated old heap that couldn't exceed 15 m.p.h. No bother. Thursday morning's post brought a letter from the nice man, however, which revealed that the "few invited journalists" would be competing against Derek Bell (F3 driver, second Grovewood Award this year), Mac Daghorn (F3 driver—you must have heard of him) and Peter Westbury (F3 driver, former British hill climb champion, etc.).

Which is how this luckless scribe (several rallies, driving tests, etc., but a couple of years ago) found himself pitted against about 12 other *selected* journalists plus the members of the F.I.R.S.T. F3 racing team, and had to try damned hard not to be last.

Now to the Mokes themselves. Four of them were built at the beginning of the year by Felday Engineering Ltd., of Forest Green, Surrey (Peter Westbury's firm), and have been used for demonstrations and as additional attractions at various rounds of the Player's No. 6 Autocross Championship throughout England all season. They use the Moke body, fitted with a roll-over bar, bucket seat and small diameter leather-covered steering wheel. The real modifications are under the bonnet, where a 1,275 c.c. Cooper S engine, with a Janspeed competition manifold and 510 camshaft sits. The engine is Brabham-balanced, and has a light-weight flywheel, and the other touches are straight-cut close-ratio gears and a 3.7:1 final drive.

All this makes what I always considered a deb's kinky round-town transporter into a somewhat hairy vehicle. Anyway, the Mokes, the journalists and the racing drivers assembled on the Friday morning in a field which looked very nearly as muddy as the High Wycombe championship course —and that's muddy.

The nice man pointed at a Moke, and said: "You can go first. Have one warming-up lap to see which way the course goes, and stop on the line—then you have three timed laps". I climbed over the roll bar and

Trying to pass on the outside, and throwing plenty of mud about in the process.

Why Maurice Rowe has given up sitting on the outside of corners. The author doing a "phenomenal avoidance".

How it should be done. Mac Daghorn grits his teeth and winds the wheel, with his right foot firmly down.

into the bucket seat, and then had to get out again because I was sitting on the full-harness safety belts. These were fastened round me, and I found I could just get the clutch to the floor, had fine control over the steering wheel, but needed a six-inch extension to the left arm to change gear easily. They said not to worry about that: "Into second as soon as possible and leave it there".

Heavy rain overnight and probably for the past few days had left the area of the starting line rather squelchy, and the remainder of the course, where the Felday mechanics had been having a little practice before we arrived, was worse. As soon as the power was turned on it seemed virtually impossible to keep the thing straight, and the practice session was spent going energetically from lock to lock and reflecting on the incredible amount of mud which tiny wheels could throw up.

Then it was up to the starting line, with Ken Best waving a white handkerchief and wearing a happy (or was it a mocking?) smile. Down came the flag, and off! Remember what the man said, into second as soon as poss. That's it, stick to the dry bit for a little more adhesion. Into the first corner, a right angled one. Right into the ruts, opposite lock, a bit more right foot —blast, shouldn't have lifted off. Now the chicane, full right lock, then left lock, then straighten slightly. There's Maurice Rowe—bet he wouldn't stand there if he knew how little control I have over this thing. Round the next right-angle (the course was roughly square), full bore up the straight, try not to lift off for the corner, on to the dry stuff, and up to the next one. This corner's a bit tricky—slight adverse camber and a bit more than a right-angle—but don't lift off till the last minute, do the Highland fling on the wheel, and we're heading for Ken Best again.

Somewhere about this time realization dawned fully. The quickest way round an autocross course is to keep the beastie sliding as much as possible. After this is appreciated it's only a matter of nerve to see if you can go all the way round all the corners without once lifting off (somebody actually *braked* once). After the three laps, the time was recorded as 2 m. 45.4 s., and then the sun came out, curse it, and almost everybody started going faster as the track dried out.

To crown it all Derek Bell jumped into a Moke, not long before my second attempt was due, and went round at what sounded like a steady 6,000 r.p.m., without even twitching his big toe. His time was the best of the day—2 m. 15 s.

Then came *Motor's* second run, which started well, with the nerve holding out better on the corners, and the wheels spinning less on the straights. As the Moke came back to the line for the first time, another one was started off, and I couldn't get to the first corner before him. And that really was that. I spent another two laps trying to overtake, but succeeding only in getting mud plastered from head to foot. Significant, perhaps, that my second time was 2 m. 32.0 s., and his was 2 m. 36.0 s. I was baulked!

And then, just to rub salt in the wound, blow me if the nice man didn't send out a press release listing all the times. But I enjoyed it, anyway. . . .

M

Farmer Moke at a journalists' ploughing competition. One of the surprising things about autocross is that soft-pedalling doesn't give as much acceleration as flat-out (6,500 rpm) wheelspin

MINI-COOPER S MOKE

Player's No. 6 Autocross Mini-Moke

Moke comes to town. The Barton Accessories Fibreglass Sedan is neat and practical. This is the same car as above, actually the "Southern" one prepared by Felday and still wearing its autocross SP44 tyres

THANKS to the generosity of John Player and Sons Ltd., quite a few motoring scribblers got a taste of autocross driving last season at the wheel of one of the four Player's No. 6 1275 Mini-Mokes. *Autocar* was allowed to drive the Felday-prepared one on a wet drag strip in June (Santa Pod), round a muddy field in October and—after conversion to relatively licensed respectability as a neat but severe boulevard hardtop—on ordinary roads in January.

If you look up the word in the Oxford Dictionary, it says "Moke, n. (slang). Donkey [?]."
The [?] bit means that not one of the droves of dons who made that mighty work know where Moke came from, which is rather how some of the more elderly and reckful members of *Autocar*'s staff felt after they'd driven the road version. A few of those even declared that they did not want to know, but others who'd had a go on grass were keen to find out what gave this donkey its mulish kick.

The ordinary Mini-Moke is a cross-country version of the familiar BMC Mini, a poor man's pseudo-semi-Land Rover first commissioned by the armed forces five years ago. It has many uses on the more inaccessible parts of farm, moor, forest, mountain, swamp and South Kensington, where it is ideal for being youthful, windblown, conspicuous and fashionably photographed in—or rather on. As a Player's No. 6 autocross special, each Moke was stripped of its windscreen and headlamps and fitted with a Brabham-balanced stage 2 1275 Cooper S engine with twin $1\frac{1}{2}$in. SU carburettors, "510" camshaft, a Janspeed competition manifold and an unsilenced exhaust. So that drivers stayed on its back when Moke bucked, a well fitting racing howdah with full harness was provided. To preserve headroom if Moke laid down to scratch its back, a John Aley Roll-Over Bar stuck up behind, looking like chromium plated steel gamekeeper's larder to hang careless dead journalists from as a warning to the others. It had close ratio gears, a lower final drive knobbly SP44 tyres, and a leather-rimmed steering wheel.

You climbed into the seat thinking that the driving position was designed especially for well-developed chimpanzee bus drivers. Someone had hopefully put a 2in. extension on the gearlever but my arms were still 4in. too short if I sat back. Which I did after selecting 1st on the start line for three timed laps of a flag-marked circuit round a field. But the way the little beast leapt forward despite furious mud-slinging wheelspin immediately made me sit up again. Putting it into 2nd and forgetting all further gearchanging was fine and suited my limited powers of concentration.

Without wasting space on many mistakes, eventually found that cornering was a matter of pointing the two front wheels in the right direction at the right time without worrying greatly about the back wheels. Down the straights you kept your foot down flat, with the rev-counter saying around 6,000 which the engine and exhaust noise insisted was a gross underestimate while you searched urgently for grippier grass. Some way before a corner you twitched the wheel hard into it and the car went sideways, slowing somewhat and rounding the bend in great banging boneshaking bumps, greater, that is, than the ones down the straight. The flying was pretty level if not straight, thanks to Koni-damped Mini suspension (non-Hydrolastic). Opposite lock was instinctive but wasted time and space—it seemed better to keep turning further in slightly though not too much. Mud and grass flew up perpetually. It was tremendous fun. The final proof for me that back wheels on autocross Mini-Mokes are irrelevant came as I stopped and someone asked what effect I'd felt when the left-hand rear suspension arm bracket had broken; the wheel had about 15deg negative camber and was leaning against the body. I hadn't noticed.

Our first meeting had been at Santa Pod Raceway on a wet Saturday in June fairly early in No. 6S's career; it performed magnificently, doing a standing quarter-mile in 17.3 sec. As the performance figures show, 37 autocross events plus the addition of a large and roomy hardtop plus two heavy *Autocar* testers and test gear made a big difference. Barton Accessories of Plymouth make a great range of bodies and improvements for Mokes; the hardtop conversion is their Fibreglass Sedan with sliding windows. A rear bench seat and another racer front seat had been added and so had a silencer, of sorts. The body is neat, weatherproof, and attracted a surprising amount of attention. Except for big blind-spots at each windscreen pillar, and an incurable tendency to mist up, visibility is excellent, making traffic driving very easy.

On a legitimate Mini-Moke such a body is an excellent aid towards domesticating the animal. With a 1275 S engine which likes to be revved up to 6,000 rpm and gears which give a fine straight-cut whine, the body could hardly be blamed for emphasizing such uncouth additions. Conversations about the passing scene at anything above 40 mph would have been helped by a headphone intercom set from, say, a government surplus tank. It was better to journey alone

PERFORMANCE DATA
Strong wind (20-30 mph), raining

Figures in brackets are for the Morris Mini-Cooper S tested in AUTOCAR of 14 August 1964.

Acceleration Times (mean): Speed range, gear ratios and time in seconds:

mph	Top		3rd		2nd		1st	
		(3.44)		(4.67)		(6.60)		(11.02)
10-30	9.6	(8.2)	6.8	(5.5)	4.6	(4.0)	3.9	(2.9)
20-40	7.9	(7.5)	6.2	(5.4)	4.1	(3.8)	—	
30-50	8.9	(7.5)	6.4	(5.4)	4.8	(4.4)	—	
40-60	9.9	(8.3)	6.9	(6.0)	—		—	
50-70	9.6	(9.4)	8.4	(7.5)	—		—	
60-80	14.0	(12.3)	—		—		—	
70-90	20.7	(17.5)	—		—		—	

FROM REST THROUGH GEARS TO:
- 30 mph .... **4.1 sec** (3.5 sec)
- 40 mph .... **6.3 sec** (6.0 sec)
- 50 mph .... **8.5 sec** (8.2 sec)
- 60 mph .... **12.8 sec** (11.2 sec)
- 70 mph .... **17.5 sec** (15.4 sec)
- 80 mph .... **27.0 sec** (23.4 sec)
- 90 mph .... **39.6 sec** (34.7 sec)

Standing quarter-mile 18.9 sec (18.4 sec)
Standing kilometre 35.7 sec (—)

MAXIMUM SPEEDS IN GEARS:

	mph	kph
Top (mean)	91 (96)	147 (155)
(best)	96 (98)	155 (158)
3rd	75 (74)	121 (119)
2nd	56 (54)	90 (87)
1st	37 (33)	60 (53)

OVERALL FUEL CONSUMPTION FOR 576 miles: 23.9 mpg (28.5 mpg)

for any distance over 10 miles otherwise one arrived hoarse. The high vertical sides caught side winds which deflected the car noticeably.

The Player's No. 6 Moke is a noisy little beast. Our acceleration runs started with a yawp and a yell from the tyres as the clutch bit at 5,000 rpm. Then there was a shrill rising scream from the gears and a harsh clatter from the engine. Moke was trying hard and wished everyone to know. At 60 mph the engine hatch (it's too basic to be dignified as a bonnet) began to lift and tug at its straps. At 70 we were darting playfully towards the fences each side (strong side wind). At 80 it looked like a speeded-up sequence from a Keystone Kops film involving a toy fire tender, two mad scientists (us) and a mono-cyclist (we had our fifth wheel behind). At 90 the single wiper blade went into a fit, frantically beating at the windscreen trying to get into the cab and eventually giving up by throwing itself over Moke's shoulder, never to be seen again. All in all, a vehicle best enjoyed by people who find true peace of mind, which will probably pass all understanding, in such sublime things as hysterical arguments with parking wardens, getting out and shaking tardy traffic lights, or just plain punch-ups.

Michael Scarlett ☐

Left: 1275 c.c. engine fits well where it's meant to and has forced a fit where it wasn't (see carburettor trumpet clearance). Right: Our photographers spend a lot of time looking for appropriate backgrounds for cars of character

Here come de Coke Moke
Coke adds life!

A 1275cc Cooper S engine, front discs, Paddy Hopkirk rally seats and Sunraysia wheels certainly add life to the standard Moke. Such a car is the Coke Moke...

Pick a vehicle to compete in a gruelling rally from London to Sydney.

Maybe the choice would be a rugged 4WD. Or a big, strong conventional car. But a Mini Moke? That vehicle would probably be most people's last choice.

But, undeterred, Danish born adventurer Hans Tholstrup (who is not renowned for doing conventional things) and former magazine editor and ex-DJ John Crawford chose a diminutive Moke for their assault on the world's most gruelling motoring event — the 1977 Singapore Airlines London to Sydney rally.

Held over 30 days on a route covering more than 30,000 km, the Singapore Airlines sponsored event took drivers over road conditions varying from the Great Salt Desert of Iran and the Gibson's Desert of Australia to the snowy Australian Alps. Still, Tholstrup and Crawford chose a Moke. And Leyland backed them all the way. It was an excellent chance to prove that the Moke is a durable and reliable machine — despite its diminutive size.

In reality the Moke never had a chance of victory. It was simply not fast enough. Crawford and Tholstrup had to finish though. That was their ambition. If they could finish such a gruelling rally, it would indeed be a great advertisement for the Moke.

And finish they did. They were hundreds of hours in penalty points behind the rally winner, Andrew Cowan in a Mercedes. And they missed many of the gruelling special stages, when the drivers had to go quickly over difficult terrain. But they finished — in 35th place. And that was all Leyland wanted.

Coca Cola sponsored the Moke. And the little machine became known as the Coke Moke.

The body of the rally Moke was double spot welded in all the critical places for extra strength. A Mini Cooper S 1275cc engine was fitted in place of the conventional and distinctly under-powered one litre four-pot unit. The rally Moke developed a healthy 75 bhp.

Disc brakes were fitted at the front. A Cooper S sub frame and transaxle unit was fitted. English made Paddy Hopkirk rally seats were used. The Paddy Hopkirk seats are very similar to the Recaros or Scheels.

"However the Recaros and Scheels wouldn't fit in the Moke. They were too wide," said Crawford. The Paddy Hopkirks just fit in.

Four Marchal driving lights and an aluminium roo bar were fitted at the front. Sunraysia wheels and Olympic WT (Winter Tread) radial 175 R 13 tyres were used.

The suspension was standard, except that heavy duty Koni shockers were fitted. Also, English manufactured adjustable suspension trumpets were used. These adjustable trumpets meant that the rubber cones in the suspension could be compressed, giving adjustable ride height. The only other modification, apart from a heavily changed dash, a slightly different soft top, full seat belt harnesses, a roll over bar and an extra fuel tank, was English built engine stays to resist the torque reaction of the Cooper S engine.

"We had few problems on the rally, but unfortunately the two major ones we had were both very time consuming," said Crawford.

"In Yugoslavia, our English built engine stays broke. The engine twisted and cracked the inlet manifold. And that cost us a lot of time." Unfortunately the crew weren't carrying any spare stays. "And getting parts for a Mini Moke in Yugoslavia isn't easy," Crawford said.

"Our other major problem was with the English built adjustable suspension trumpets. We snapped a trumpet in Cleary in Western Australia. And again that cost us loads of time and many penalty points.

"The other problem we had was that we kept breaking the lower shock absorber mounting bolt. This was because the stroke of the Koni shocks wasn't as long as the wheel travel. We broke six lower shock mounting bolts in all, before we fitted standard Moke shocks in Alice Springs. The Moke shocks don't last nearly as long as the heavy duty Konis, but at least we stopped breaking the mounting bolts. Really though this problem didn't cost us much time. We were replacing the broken mounting bolts very quickly." And those were the only mechanical problems the Moke had.

"In other words we didn't have a single failure with a standard part," Crawford added.

Bushdriver did a lengthy test of the Coke Moke in the condition it finished the rally in.

The vehicle was in remarkably good condition after such a gruelling rally. There was a slight rattle in the exhaust, but other than that the vehicle felt and sounded very tight.

The Moke started first time every time — with one exception. After we washed the car with a hose, water must have got on the ignition. A quick spray of WD40 solved the problem. But after only a light hosing we began to wonder what the Moke would have been like after a

heavy rain storm or passing through a creek crossing (as was encountered on the rally).

Minis have a bad reputation for susceptability to damp ignitions. And the Coke Moke was no exception.

The soft top had done the Adelaide to Sydney section of the rally. The zippers of the London to Adelaide top were stuffed. They were full of dust and inoperable.

The soft top had no clear rear three-quarter windows. Other than that it was standard.

When driving the Coke Moke the first impression is that the rally Moke is considerably better than the standard machine. The Cooper S motor makes an enormous difference. And so do the rally seats.

If Leyland put their 1275 donk (especially in good ol' Cooper S trim) back into the Moke, the car would be an enormous improvement. The Cooper S unit is lively, spirited and sounds unreal. With it fitted, the Moke is a great fun car with good acceleration and all round lively performance.

The Paddy Hopkirk seats are extremely comfortable and are incomparable to the standard Moke 'park bench seats with a bit of foam'. With a lively motor and comfortable and correctly supporting seats, the Moke is a much, much better car. And with a full roll over bar and full harnesses, the Coke Moke is also a considerably safer machine than its standard brother. Not that I'd like to have a prang in any Moke (especially in the side, where the Moke offers little more protection than a motor bike).

The gearbox is standard. Even after 30,000 km of London to Sydney motoring, the gear change was precise and easy. It is a great improvement over early Minis.

The brakes were disappointing. The brake travel was abnormally long. And the brakes never really gave confidence. They never felt strong.

When Bushdriver drove the Coke Moke it was fitted with standard shockers, as it had been in the rally since Alice Springs. The car handled well. On dirt the Moke was extremely easy to control and very predictable in its handling. It didn't cope with corrugations well, however. On corrugations or in rough conditions there was tremendous feedback through the steering wheel — enough to break your wrists if you maintained a firm grip on the wheel. The Moke wouldn't ride the corrugations either. Rather, it hopped and pattered over them and became unstable.

On bitumen the handling was also generally impressive. The front Olympic tyres were badly worn (they were the original tyres fitted for the start at London and did the whole rally) and screamed at the slightest provocation.

Even when cornering gently the tyres would shriek, giving a false impression of the speed of cornering.

By contrast the rear tyres (one had done the whole rally, the other was fitted after a flat early in the event) both had plenty of tread. They looked as though they'd happily do another 30,000 km of London to Sydney rallying.

Bushdriver was impressed with the Olympics. Despite the badly worn fronts the Moke's handling, particularly on the dirt where the worn tread should be most noticeable, was good. And even on the bitumen (a surface that winter tread tyres hardly shine on) the handling was impressive, despite the early screams from the front tyres in particular.

Like Minis and other front wheel drive cars the Moke tends to understeer (or has front end slide) on both bitumen and dirt. By flicking the car, or setting it up, before the bend the characteristic can be overcome on the dirt.

The dash of the Coke Moke was heavily modified. Directly in front of the driver is a newly added panel that houses the extra fuel tank level gauge, a tachometer, water temperature gauge and warning light, oil pressure gauge and ammeter.

We weren't impressed with the layout of this special rally panel. The two most obvious and easy to read gauges for the driver were the extra fuel tank level and the water temperature. However the tacho and oil pressure gauges were partially hidden by the Prototipo rally wheel.

Despite the 30,000 km ordeal the Moke had taken the oil pressure and water temperature remained healthy throughout the test. The big yellow water temperature warning light is directly in the line of not only the driver's vision, but also the sunlight. Consequently with the sun shining directly on the light you can't tell if it's on or off.

To the right of the driver is another added panel, mounted on top of the right hand side box section, that contains switches for the driving lights and a special and novel siren, used for dispersing the crowds in heavily populated countries like India and Pakistan.

To start the Moke you turn the ignition key and then press the starter motor button, just to the right of the key.

In the central console there's the standard Moke speedo (to 150 km/h) with standard Moke fuel tank level gauge housed in it. The gauge didn't work properly. It showed half full when it was actually full and empty when it was between half and empty. The intrepid Bushdriver test team actually ran out of fuel climbing the Mt Victoria Pass in NSW. And that wasn't much fun...

Also in the central console there's the windscreen washer switch, the headlight switch and the wiper knob and choke. The windscreen wiper switch didn't work. Nor did the washer switch or the choke. The heater demister ventilation control is directly under the central console.

In front of the navigator is the Halda Twinmaster odometer — which is standard wear on all rally cars — plus a compass, a fusebox, a terribly inaccurate clock which lost 1½ hours every four hours (!) (as one whacka said: "No wonder the Moke was late for so many controls if they went by that clock"), a map light switch and an interior light switch.

Around town the Coke Moke is an extremely nippy (no pun intended at the Japs) little vehicle with plenty of zip and zap.

The basic Moke is a fun car. But if it was like the Coke Moke it would be a tremendous fun car, because the rally Moke was enormous fun to drive.

Leyland has long been criticised for the shortcomings of the Moke.

And yet in the Coke Moke Leyland has a vehicle that is vastly improved, with so many of the faults cured.

Hopefully the Leyland boys will learn from the Coke Moke. And hopefully Mokes of the future will be more like the Coke Moke than the current Moke of today.

John Crawford summed up the Coke Moke well.

"It is a tremendous fun vehicle that goes very well. But," and he paused, "can you imagine driving from London to Sydney in it! I still have nightmares just thinking about it."

The Coke Moke is a great fun car. But in reality it is not a viable alternative to the Jumbo jet in London to Sydney travel! ♣

WE MADE IT. DESPITE THE ODDS.

When we entered our Leyland Moke in the London to Sydney Rally, the odds were stacked high against us. The experts told us we wouldn't make it. One month and 30,000 rugged kilometres later, the drivers, John Crawford and Hans Tholstrup showed everybody what a tough, rugged go anywhere machine Moke is. Perhaps the only thing more incredible than what a Moke can go through, is what it goes for.

STANDARD MOKE: $3125
MOKE CALIFORNIAN: $3729

Recommended retail prices exclude registration, insurance and a delivery fee of approx. $75 which may vary from dealer to dealer and state to state.

1987

Vol. IV No. 3

THE MOKE CHRONICLES

moke \'mōk\ *n* [origin unknown] **1** *slang Brit* : DONKEY **2** *slang Austral* : NAG

RaceMoke

The Moke pictured here is owned by <u>Rear Admiral & Mrs. Robert A. Phillips, of Mechanicsburg, PA.</u> Robert writes:

In late 1970 I acquired good old 895551 from the Budget Rent-A-Car people in Honolulu after someone tried driving it without any radiator water, hence the price of $115, was about right. I then performed a heart transplant with a Mini Cooper 1275 S donor. I then added Cooper disc brakes, altered suspension pieces, constructed a nice rigid roll cage, made custom fitted seats by sitting in Saran Wrap over fiberglass cloth & resin which had not yet set up, & topped it off by a new dash panel made from my wife's cookie tin taken when she wasn't watching.

The super wide wheels & tires didn't arrive until a few years later in Pennsylvania, & they gave the car a definite visual relationship to the then popular cars for kids called Hot Wheels.

It has been an absolutely incredible car to drive. The NHRA timers at Hawaii Raceway Park certified a 15.27 elasped time for the quarter mile. In 21 autocrosses & 10 Pennsylvania Hillclimbs it has earned 16 1st Place (including 5 fastest time of day runs), 11 2nd Place, 1 3rd & 3 4th Place finishes.

The slightly more aerodynamic front end treatment happened after the July 14, 1974 Giants Despair Hillclimb during which I demonstrated the value of six point seat belt/shoulder harnesses when using a snow fence & earthen embankment to stop from about 40 miles per hour in about four feet. The test is <u>not</u> to be recommended if you have any regard for your

neck at all.

The four carburetor stacks were very scientifically designed by hammering a towing bar ball into the end of some copper sink drain tubing to make a flair, & then finding a chrome shop.

With less than $1200 invested it has proven to be a very good value for the money, more so if I were able to melt down some of these dusty trophies.

As most other Mini owners, I seem to have accumulated a few rare & highly valuable surplus parts. Should any of your readers have a need for any of the following they should contact me for heavy price bargaining & shipment instructions:

1 MG100 eng., transaxle, remote shifter unit

1 Complete front sub frame, including dry suspension & brakes

2 Mini steering racks, including tie rods

3 Narrow ten inch rims

2 Stock Mini 850 cams

I trust you find the story & pictures of 895551 a bit different from the ordinary, but then who ever said anything about a Moke was ordinary?

The Cult Moke

The Mini Moke was launched at the start of what became known as the swinging sixties and it became a symbol of that era. As it was a very non-conformist car it fitted in well with the carefree attitude of the period and became popular with the fashion seekers.

The Moke starred in many films in the 1960's, and still does when a sixties feel is needed. It played a role in the Beatles' film *Help*, and was used by the enemy inside a volcano in the James Bond film *The Man With The Golden Gun*. Perhaps its most famous part however, was in *The Prisoner*, Patrick McGoohan's fantasy series, where they were used within the village as taxi's. There were four or five Mokes used in this series. The special canopies, seat covers, wood grain panels etc. (see photos over and below) were done by Wood and Picket of London in 1966. One of these Mokes is now owned by the Prisoner Information Centre in the village of Portmeirion, North Wales, where the series was filmed.

The Moke was also used for advertising, its frame being easily adaptable, and BMC had a parade team of about 15 Mokes which were loaned out for events such as the Lord Mayor's parade, golf tournaments, and motor racing events, where they were used for people movers, advertising display vehicles or even for carrying a jazz band.

Photos by Max Hora

DEMOBBED!

Drummed out of the armed forces for not being tough enough, the Mini Moke found its place in life among the beautiful people of the sixties, as Giles Chapman recalls

Mokes on parade in McGoohan's Prisoner *TV series*

What epitomises the late sixties? Perhaps it's songs like Donovan's *Googoobarabagagel*, maybe *The Avengers* and movies like *Having a Smashing Time*, possibly Mick Jagger's Mars Bars ... Mods on the rampage in Brighton ... hippies telling you to love one another ... wide ties ... sideburns ... cars ...

Cars. And then there was the Mini Moke – the world's first true non-car. It drove and steered and carried people and used petrol, but it still wasn't really a car. Now, the Jaguar E-type was a car, and so were the Hillman Super Minx and Vauxhall Cresta. You sat in them and drove about in order to achieve sensible aims: going to work, a drive in the country, shopping, transporting elderly relations ... *keeping warm!*

All these things may just have been possible in the Mini Moke, and yet still it did not represent what a car should.

The Mini Moke is jolly possibly the only vehicle which is almost totally a fashion accessory and film prop, and to confound matters further its intended purpose in life couldn't have been more different.

It is a very valid part of the sixties British culture that has been ignored for so many years in the face of such so-called phenomena as Levi's jeans, drive-in movies and Elvis Presley. While most American culture stems from the fact that the place is so big, most British culture comes from our perceived notion that we are very much bigger than we are – which, of course, we were when we had a huge and willingly-patriotic Empire. Consequently, although America is large, England is larger than life. Disturbances to the peaceful way of English life are headline material, and crazes and fads – soap operas, records, newspapers – take over like fever.

The Mini Moke was in the thick of the frenetic, 'anti-square' period of the sixties because it was a rebel from the ranks of motoring normality at a time when rebels of any kind were in great demand. Although you never owned a Mini Moke yourself, you always knew someone who did, or had seen one somewhere important. Every sex kitten's boyfriend had one in Chelsea or on the Portuguese Algarve, and they could be relied upon to be parked in odd places like in restaurants or on houseboats.

The Mini Moke was not the car its makers had intended. BMC was about as 'square' a company as you could find. Vast, corporate, very conservative, they made Austin Cambridges and bakers' delivery vans. However, they *did* make the Mini, a truly inspired design with a front-wheel drive/transverse-engined package that set the scene for the modern and nowadays ordinary small car. Like all companies, BMC tried to explore all ways of exploiting this design to get maximum manufacturing use from it. It wasn't long, therefore, before there was an estate, a van, a pick-up, even small luxury saloons and sporting derivatives, and the Mini went into licensed-production in places as far apart as Italy, Australia, Spain and Belgium.

Aside from such civilian markets, there might also be business, thought BMC, to be gained from supplying the armed forces. This was an area where BMC was rapidly losing out to Rover's Land Rover, since the Corporation's own Austin Champ was decidedly old and unwieldy, and its new Gipsy, although a solid piece of engineering, offered little more than the Solihull product – added to which it was made of steel and not aluminium, and therefore had rather more limited life expectancy.

Unable to compete head-on with Land Rover, BMC turned their attention to the Mini concept – of which enough has been written not to warrant a detailed description here – to see if any of its talents could be harnessed to provide military transport. Although the new vehicle would have only front-wheel drive, it was thought that it could be light enough for four fit soldiers to carry over very rough terrain. It sounds crazy and it was, for although an unladen car would weight 2.5cwt, the addition of any sort of luggage meant that four pairs of shoulders

Moke in its natural habitat – warm foreign climate, pretty girls, quayside yachts – a far cry from the army mud-plugger that BMC had originally envisaged ...

would be very weary after only a short distance.

The army duly agreed to take the car on trials to evaluate it with a possible view to placing orders. The first prototypes began tests in October 1960. Naturally, it was expected that having the engine weight over the front wheels would give excellent traction, but once again, with four beefy soldiers aboard, the wheels became too lightly loaded, and a clean take-off on a slippery incline became difficult.

The next manoeuvre was to see how the car would behave when parachuted from an aeroplane. It was light enough for use, but having all the aerodynamic properties of a sideboard, airborne descent was hazardous. Testing, however, continued apace, and the car was shown for the first time to the public at the Amsterdam commercial vehicle show in February 1962. It was christened the Mini Moke, a moke being the Australian slang term for a very poor horse! The engine size was increased from 848 to 948cc, so that the car did not lose performance when running on a low compression and using low-grade fuel. The Moke used its own special floor pressing, with sub-frames at the front and rear carrying the independent Moulton rubber cone suspension, and ground clearance was 6ins, so that at least the squaddies could have good roadholding if not ultimate trans-terrain ability.

The bodywork, what little there was, consisted of simple, flat panels for the bonnet and wings, while enclosed steel panniers made up the doorless sides, over which one stepped to get in or out. As these were hollow, various pieces of kit could be stowed away in them. At the front was a blunt 'Hovis-shaped' wire-mesh grille with round headlamps either side, while the bumpers looked like small scaffolding poles secured onto fixing brackets. The spare wheel was attached to the rear and there was a tilt-type hood with no sides, so that when the hood was folded down, the car was more open than any sports car of the time, and indeed it looked as if you would sit on rather than in the Mini Moke. The 10ins wheels looked absolutely lost in their vast wheel-arches.

For once, *The Autocar* made an intuitive comment: 'The Minimoke (sic) is not intended for private ownership, though it may have some potential as a cheap, go-anywhere vehicle'.

Despite the thumbs-down from the armed forces, the BMC experimental department had not lost faith in their idea, and next they came up with a twin-engined Moke, nicknamed the Twinni. The concept was delightfully simple: there was a 948cc engine with transmission at the front, and an 848cc drivetrain mounted at the rear with steering locked in dead-centre position, a lever beside the driver providing gearchange for the engine at the back, giving him a gearlever for each hand. The two motors were controlled by single clutch and accelerator pedals, but the engines could be run separately as they each had their own ignition and starter button. The road surface was the medium that co-ordinated the different power outputs of front and rear engines ...

This incredible machine could get across slippery or damp surfaces at incredible speed, although, as the ground clearance was the same, anything marshy or very mountainous caught it out, and this was why, perhaps, most of the armed forces which evaluated it gave it the bird in favour of the Austrian – and much taller – Steyr Puch Haflinger.

Obviously, without major change to the design, which would mean abandoning the Mini base in the main, the Mini Moke was not going to be bought by the world's armies, and all developments in that direction came to a halt. It is surprising that the project got as far as it did in view of the limitations of the design in the first place, but there was to be a twist in the tale of the Moke. The two-wheel drive car was now fully developed, so in order to recoup their already considerable development costs, BMC decided to put the Moke on sale to the public.

It went on sale in August 1964 for £335 basic (£405 7s 1d with purchase tax) in pretty much the same form as it had appeared at Amsterdam two years previously. The crude wire grille had been replaced by a slatted metal air-intake at the front,

Moke in the air – aerodynamics hindered descent!

and the headlamps were set closer together. But the most important change was the reversion to an 848cc engine, perhaps demonstrating BMC's wish that it should *not* be used for arduous off-road activities. The unitary steel body was fully rust-proofed and the canvas hood-cum-tilt was sand-treated and had detachable support tubes. The windscreen folded flat and could be removed.

Two identical versions were offered in normal BMC style – an Austin and a Morris – and the car was intended as a general purpose runabout, such as a beach car, hotel taxi or factory transport. In its basic state, there was one (driver's) seat only, with the others offered as optional extras, and thus it was hoped that it could be sold as a commercial vehicle rather than a car, avoiding purchase tax enforcement. This seemed fair enough, but the Customs and Excise office did not agree and insisted that the tax must be paid even on one-seater versions. At once, the price advantage that might have sold this vehicle against more expensive competition was removed, and the Mini Moke's fate was sealed – it was a car, and a car that only an eccentric would want to own. Love it or hate it.

You really had to take the Moke in the spirit in which it now found itself. *Autocar* found this difficult: 'First and foremost a cross-country vehicle, it makes

Moke awoke – the original, bluff-fronted utility revealed at '62 Amsterdam show

Moke in a moat – limited ground clearance limited most military applications

Moke folk – fashionable people loved the car's cheeky London impracticality

Moke-on-the-box – Persuader Tony Curtis and amie on French Riviera, 1971

an ideal runabout for farm or estate use, and although it may tempt some to buy it for ordinary motoring, it should be remembered that this is a secondary function'.

The magazine continued: '... its compact size and unobstructed visibility make it a splendid vehicle for nipping through traffic'. They then took it into the country and came up with: 'The sort of gullies, hummocks and potholes of a cross-country course, which one would expect to make the vehicle leap into the air, can be taken very fast without any fear of heads making holes in the canopy'.

Cars and Car Conversions, in 1967, was far less stuffy: 'It's rather like World War One flying – there is no way of keeping either warm or dry except by virtue of the clothes you put on. So you put on a kit which makes you look like a cross between an intrepid aviator and a motorcycling fly-fisherman, climb up onto the driver's perch and wind up feeling like Nanook of the North without his Kayak. The Moke has a hood, but it doesn't matter much whether you put it up or not. With the hood up, no rain gets inside unless it is horizontal-type driving rain, or unless the Moke is moving. With the hood down, no rain gets inside if the Moke is moving, unless it stops.

'There is something more exciting than driving the Moke – it's being a passenger in one. There comes a point, on all corners which leave the left-hand side of the car on the outside, when you realise that there is nothing between you and the ground whistling past underneath at up to 60mph except God's good fresh air. Which may have excellent remedial qualities, but we never heard it was much good for a broken neck.' Their sum-up was that the Moke was great fun if you didn't treat it like a tractor!

The Moke stayed in production for four years at Longbridge, and of the 14,518 made, over 90 per cent were exported to lands where purchase tax was a funny foreign saying, and the sun shone all day long. They were particularly popular in the South of France, Italy and cosmopolitan cities like Paris, where it was possible to hire one in 1967 at hourly rates, so that you could have a panoramic sightseeing tour of Paris on your own. A few hundred stayed in London, where they were the grooviest thing for boutique-ing and drunken drives home after parties. The age of the mini-skirt was all but over when the Moke became available, but the new peace-craving hippies and beautiful people found them just the vehicle for their cause, painting flowers on them or psychedelic blobs and abstracts that matched the exaggerated shirts and foot-length dresses of the period. Maybe it was also the fact that the Moke had been rejected by the armed forces that increased its appeal to the peace lovers ...?

One export market that looked especially promising was Australia, and in 1966 BMC announced that a special version would be made out there at BMC's Victoria Park, Sydney plant. Known simply as the BMC Moke, it had 13ins wheels giving a couple of extra inches ground clearance, a reworked, beefed-up suspension and all sorts of tough bits like bigger mudguards, mudflaps and stone guards. It proved very popular, building on the already-established cult status of the Mini Down Under, and when BMC finally gave up on the car that had given them so many problems in the UK, they transferred all Moke manufacture to Australia, where it continued to be built next to Leyland trucks and Land Rovers up until 1982. At this time, BL were reconstructing their overseas operations and the Moke transferred yet again to Portugal, where it continues to be built today by a subsidiary of Austin Rover called IMA in Setubal. A press release issued by Austin Rover in 1984 said 'The availability of the Group's world-wide sales network will increase this unique vehicle's export potential', but so far, this prophesy, far from being fulfilled, has largely been forgotten. In the autumn of 1983, Dutton announced that they would be the new concessionaire for the UK, but only managed to sell about 20 here before the Portuguese decided it wasn't worth their while.

The Mini Moke had had a short and troubled life in

Mokeability – the only way to see Paris in 1967

Britain, but this did not stop it endearing itself to the hearts of the British public. Rather like the Morris Minor Convertible, it became one of the most popular 'character part' cars on the big and small screens. Most famous of all its television appearances was in *The Prisoner*, Patrick McGoohan's fantasy series about a secret village where intelligence agents are kept against their wills until they confess as to why they've resigned, and of Number Six's attempts to resist and eventually escape. The bizarre nature of this programme and, some say, its sinister implications of real life, suited the zany character of the Moke, four or five of which were used as village taxis. The underground scenes in Roger Moore's Bond movie *The Man With The Golden Gun*, contain several bright orange Mini Mokes, property of the vilainous enemy, of course, and the Moke played a part in everything from *The Avengers* and *The Persuaders* (*that damned programme again – Ed*) to the children's TV show *Tiny The Giraffe* and the feature film *Crossplot*.

You still occasionally see a Moke about, generally in scruffy condition and driven by last-chance trendies or Twisleton-Wykeham-Fiennes types, so it would be well worth snapping up this odd little piece of British culture while you can.

Moke-on-a-rope – engineer Jack Daniels demonstrates the Twinni's capabilities

Moke with fire – two Mini power packs gave the Twinni very good traction

Moke snow-show – a Twinni clears the slush from the Longbridge lawns in 1963

Moke for wimps – optional (and substantial) hard top kept out the elements ...

MINI MOKE

BMC PARADE TEAM MINI MOKES HAVE BEEN in great demand recently, and these pictures highlight just two of their many tasks—pleasurable tasks we might add.

The main attraction as far as the writer was concerned was the Lord Mayor's Parade through the City of London. The Mokes, decked out in absolute splendour, made a great impression on the crowd. Being one of the drivers on this great occasion the writer is able to verify this claim with some authenticity.

Starting early in the morning, the Mokes were checked and fitted with loudspeakers, they then proceeded to a special pick-up point under Police escort. Here we met Kenny Ball and his

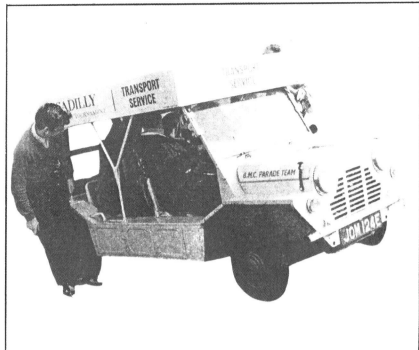

Jazzmen who were to ride on the Mokes. In addition we found we had some extra passengers in the shape of the Miss World Contestants for 1967.

During the parade the Mokes were driven in close formation and they behaved extremely well. Some of the R.A.F. flying teams would have been hard put to, trying the same stunt . . .

The other outing of these versatile vehicles featured here was on the occasion of the last Piccadilly Golf Tournament held at the Wentworth Golf Club in Surrey. Although the writer did not attend this, he understands that the vehicles came in for a great amount of praise for their work as a mobile film unit platform and transport for those in attendance.

MOKING AROUND

by
John Stanley

MICK CROSSED EARL'S COURT ROAD IN A flourish of bell-bottoms and was consumed by the late-night supermarket. It was crying outside, but who wants the hood up? Three birds sauntered up in tight trousers and tied-up shirts.

Eastern Girl: 'Hey Mister, you going our way?'

Me (*aside*): 'I don't believe it—which way is that?'

2nd Girl: 'Clarendon Road, Holland Park.'

Me: 'Would you believe, two roads from me? Mmm, yes, if you can all get in!'

The Moke screams.

Mick is helped over the roadway by a jar of Nescafé.

It's still crying, but who needs the roof?

The wind and rain dance round the tiny white machine; slithers of terraced London cross the wing mirrors; the lights turn mauve; I fancy the small Eastern one—that is, until minutes later Mick starts up conversation and we find that the three prove to be a mixture of sexes!!

It would appear that driving an open Mini-Moke in London automatically brands you as part of that rather doubtful set, 'Swinging London'. Occasionally you pass another such example, usually painted in bright colours and laden with attractive birds. Perhaps with Mick that night, people must have thought along those lines with us—who knows?

The Editor's orders were to take the BMC's Mini-Moke for a week and expose it to what he called my 'swinging' way of life and to recount it to you. Well, what can I do? Sure I meet the people and once in a while go to the right places, but heaven forbid I am classified as one of 'Them'! Forget my hair, clothes, profession; my Utopia is silent countryside, period furniture, and definitely no telly. Anyway, for the sake of a fee and a week's drive in the little thing, I agreed to try and recount how Moke and I managed. BMC themselves put the mockers on things the week prior to this, by giving their Press Moke to the Navy for a year and being quite unable to find another example anywhere in the factories or showrooms. Happiness came with a phone call from Stewart & Ardern who had a new white demonstrator. Good show S. & A., move four places forward, don't take a card, do not miss a turn.

Moke Day One was very bright: it made dreaming easy and driving an open Moke very special. Each time I drew up at a traffic jam, scores of hot, sweltering brows would wrinkle up from the shadows of their murky saloons. It's a very exposed feeling, not unlike riding a horse. I returned to my flat, where Lynn was diligently re-typing some of my work in the study. Alan Freeman was on the phone and a private call on the other. Work intervened for a while and then both Lynn and I set out in the Moke for Wardour Street, and

the offices of my own film company, Farthing Films. I say my own film company —what I mean to say is that Spencer Davis and I are joint directors of it. Together we are involved with the production of some very big and exciting motion pictures for the cinema. It would not be in our interests to detail the names and talents which have been connected, but time and other media will keep all informed. The little car seemed very at home in the famous 'film' street and I must admit to taking great delight in stepping in and out of the thing with complete abandonment.

Evening One involved a visit to Alan Freeman to talk over a film idea which we shared. He was, however, immediately taken with the little thing and obviously had never seen one before. We strutted around it for a while and then I offered to drive him round the block. No sooner had we made two of the turns involved than he was wanting to drive the thing. After persuasion (S. & A. move to the next paragraph, do not look back, do not be annoyed) I moved over and Alan, enjoying himself immensely piloted the little machine round the side-roads of the area.

Once back, he seemed very interested in buying one to add to his stable of cars and I volunteered to get some quotations.

Day Two was hard work. Film work, consuming. I had occasion to drive the thing down to the West End with a script I had completed and which had to be delivered to my own Literary Agent. *En route* a group of young people walked up to me in the jam and threw two pieces of white paper on the floor to the passenger's side. Full of annoyance at the over-

seas impudence of people using my Moke as a rubbish bin I snatched at the offending bits whilst trying to drive off again. It proved impossible, but once stopped again I noticed that far from being waste paper they were open invitations to a new night club. Obviously Moke people are just what they wanted.

At Piccadilly someone with a Hasselblad camera took pictures of the Moke with a gaggle of us from the office on board; at four o'clock I had a meeting and during the evening explored London by Moke. A crowd of us went to Covent Garden and had tea with the porters, before moving on to the strange concrete world of London's South Bank and the Festival Hall complex. By this time it was into the early hours and silently drifting through the open buildings and roadways by Moke was quite exciting

Continued from previous page

and sometimes frightening. Indeed, I have since started to write a play around just such a set of circumstances.

Day Three was Wednesday. It was a day spent around a famous London music publishers' studios. I was, in fact, there collecting material for a recording friend of mine in Italy, but got involved with a very talented set of musicians called The Mirage who were having some of their work recorded by musical director Kirk Duncan. It seemed like a good excuse for another picture. Round and round the block again whilst the photographer darted about the traffic and pedestrians trying to snatch pictures. On one circuit three gentlemen in a vast Bentley all waved. Fancy being in such a car like that; get claustrophobia! We drew up for some final pixs, singer Scott Walker pulled up by the next meter to go into the studios and Patti Boyd left with sister Jennie. We all went back to record, under Kirk, who has been involved with the production of records and songs for names like John Walker, The Hollies, and the Tremeloes.

On the following evening I had to go to dinner with an author who is currently writing a big film for Farthing. The problem was that he lived some 30 minutes north of myself, and whether to forsake the little Moke for my Cooper 'S' or freeze. Actually, I chose the latter and really enjoyed the journey out there through the thinning London rush-hour. I have never driven anything with such a zest for living. There is such complete vision that you know and can see exactly how many inches there are to spare on each side. Many times it did frighten other traffic, most drivers wearing that resigned expression of someone waiting just for the sounds of tinkling glass and bending body panels.

The journey back is inscribed deep in my memory, for it was a quite beautiful evening and the sky was perfectly clear. I thanked my hosts and climbed in for what should have been a nice romantic drive back, but once I got going and there was no traffic to keep me going slowly I realized just how exposed I was. It was fabulous driving, but very cold. Honestly, it's twice a sports car, for you aren't covered up in the slightest.

Friday was busy and down to earth again. I had meetings with our solicitors about a contract, an actress in the afternoon, arrangements to be made for tomorrow's Dee Time, the staff wages and so on. Really, the poor little Moke was hard-worked, but unnoticed. Towards the end of the day it rained and I had cause to use the roof, which, to my surprise, went up much quicker than an M.G.'s. At two a.m. I put down the tiresome script I was reading and went to bed. In the morning the Moke would have to go back and I would be again using my Cooper. The Mini-Moke is a funny little car; one with great character. It's often nicknamed a 'fun car' but this is just what it enjoys being. During that week we seemed to laugh a lot, certainly looked fitter, and caused more attention than usual. A car for Swinging London; well... that's such a silly phrase but perhaps, in essence, correct.

THIS IS YOUR CHANCE TO jump right in with the jet set – so the ads ought to say, but don't. This is the top gear car for people on the move. It's also the coldest, the wettest, the most useless cross-country vehicle since 1898 . . . but to hell with the wide open spaces. We fell in love with it strictly as a fun about-town car. We begged Henry Manney to prolong a stay in London to photograph it. We conned the prettiest birds we know (ex-art associate Hawkins's wife Jackie, and her sister Jennifer) to come and smile. We borrowed that pvc trouser suit from a Hampstead boutique, fixed a Chelsea rendezvous for a mammoth photo session, sat back – and watched the rain pour down.

This, then, is the tale of moking the mast of a Mini Moke.

Yes, it was raining. We had a week to find out why this tent-on-wheels had become the darling of the Chelsea set. The week stretched damply away into a cumulus nimbus future. Then we noticed a smile, the first for days. Our chin went up and we gave the engine a few more revs in third before doubling down showily for the next hold-up. We collected a feeble wolf-whistle, carved up a staring taxi driver and decided we'd found most of the answer already.

In fact, completely unintentionally, the Moke has most of the properties inseparable from our much-vaunted ideal town car – far more for this special use than the Mini itself. The fact that it's not exactly a success outside the Royal Borough of Kensington and Chelsea must have come as something of a shock – a shock that apparently BMC's ad-men still haven't got over, judging from the agricultural nature of their literature.

Even in traffic, though, complete contact with the outside is by far the Moke's biggest advantage. It's a quality that would be lost with even the side-screens available from Barton Motor Company. True, you get traffic fumes blowing through your hair – but you can join in kerbside conversations. You can nip in and out (or is it on and off?) at will. Parcels can be dumped unceremoniously, friends quickly piled aboard. There's more space than in a sports car, no doors to be locked, no windows to be wound, no claustrophobic iron sheet an inch or two above your head. The scanty hood is instant up, instant down – just like an English sun.

Perhaps surprisingly, this is enough to overcome the myriad disadvantages. You can get far colder and wetter in a Moke than in any sports car. We quickly learned that we pickled if we pulled up near belching diesels, that buses made welcome shelters from searing cross-winds, that if we parked left side to kerb the camber threw water from the roof over our passengers and not ourselves . . . For £406 (£4 cheaper than a Fiat 500, £34 dearer than the barest Minivan or pick-up) the Moke is delivered in the crudest possible form. All three passenger seats are extras at £25 the set, and so are the grab handles. Girt tubular bumper bars front and rear are standard, though: they have a psychological effect all round in a butchery session – which can, in any case, be carried out with surgical precision in a vehicle where all four corners are not only visible but almost within reach. Occupants sit rather precariously on a level with the wide pontoon sides. If the seats are removed, the well forms a flat baggage area. Two moulded rubber catches hold the lift-off bonnet firmly in place and seem far more practical than the Mini set-up. On the test Moke's engine, plugs and so on were waterproofed – probably with the money saved in fitting a single windscreen wiper which is little use anyway, as the inside of the screen gets just as wet.

The only chrome is around lights and speedo. There are, if possible, even fewer knobs than on the basic Mini. The only luxury is the turn-key starter à la Cooper, which long ago dispensed with the old fingernail-saving trick of using a well-placed stiletto heel on the Mini's rubber whatsit. Lucky: stilettos are out, too. The seats are all adjustable – with a spanner. Driving comfort is a dirty word. Nonetheless, just buzzing about town can be fun for its own sake, though the usual raucous first gear graunch brings more than its fair share of kerbside ribaldry. A passenger's life is even more hair-raising, especially in the front where shortage of room ⟶

MOKE A LA MODE

it mayn't be much in the mud – but in Chelsea it's fab, gear, rave

We sent Jan Condel, CAR's kinkiest staffer, to try the Mini-jeep that flopped in the country but drew every eye in Town

➡→ makes a poor wight feel unbalanced. In anything but high summer *all* occupants need protective clothing — gloves, wind- and rainproof moking jacket, rugs and foot-warmers, as all north winds make straight for the foot well.

All hell breaks loose on the open road at anything over 58mph. The pvc hood flaps dementedly and the whole outfit shivers and shakes like an ancient biplane lumbering up for takeoff. Passing gusts of wind will lend an extra burst of speed, but passing under a bridge will as quickly rob you of five hard-earned mph. The cause is all aerodynamic as the Moke, hood and windscreen in position, is three inches higher than the Mini. In fact, although none of the ratios are altered, the little car wouldn't keep up a steady 60mph over the test track's measured quarter-mile.

We drove it round our sodden skidpan and only succeeded in filling the foot well with water while proving that its roadholding is no different from a standard Mini's, despite the optional cleat-tread tyres. Suspension, incidentally, is the old rubber cone and shock absorber setup. In the quest for action, we headed for our favourite stretch of broken *pave*. Like learning to rise to the trot, the Moke and its rider were never airborne together. But we failed to bottom the heavily guarded sump, and nothing fell off except the wiper switch.

Finally someone suggested a short session in the mud. The idea! We'd heard about the vehicle's cross-country background, but really. Beaming gaily at the camera, we plopped off the tarmac and into the mire — and stopped dead, six feet from the edge. Mud flew, wheels spun and we stayed right there. None of the schoolbook theories worked, but after a time we found we could go anywhere the Moke wanted to, crabwise in reverse if we used lots of revs.

Not for farmers, this one. But we'll remember how the Moke turned city sceptics into fans. We'll remember, too, the earthy Morgan feelings. Best of all we'll remember the effect on thrusting London drivers. A touch of the horn and rush-hour traffic would halt with polite hand signals to let us cross their bows. Why, one soaking night a car drew up as we parked and a deep base voice boomed 'Madam, we admire your courage'. Seconds later four large hands had grabbed suitcases, parcels and gub and two large umbrellas escorted us to our door. In a week we found that chivalry's not really dead. ❄

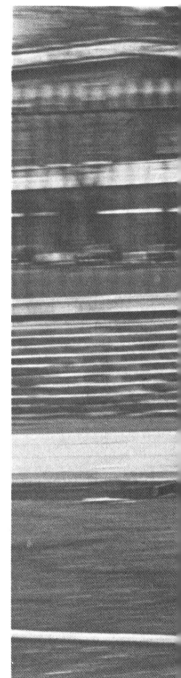

122

Windblown? So would you be (top left). Mini-Moke is more exposed than the starkest sports car, for better or worse

Styling at its simplest. BMC's newest Mini-variant has the same air of uncompromising fitness as a London bus

Room for many more. Moke's capacity for people and luggage (left) is limited only by what discomfort you can stand

Hairy performer (right). Actually, low weight gives Moke lots of acceleration although shape sadly limits its top speed

Top gear car. Reason for Moke's success with the Chelsea set (below) is its relaxed air. Hollow sides serve as boot

photographed by Henry Manney III in and around Chelsea, London

Macau Mokes

MACAU'S ORIGINAL SELF-DRIVE CAR RENTAL COMPANY

There's nowhere quite like Macau! It's in Asia and yet it's a little piece of Europe, it's Chinese and yet it's Portuguese — Portuguese colonial architecture stands side by side with traditional Chinese pagodas and temples. It's the combination of cultures and styles that make Macau so unique.

Part of Macau's charm is that it's easy to get around. The Macau Mokes Group Limited is Macau's original self-drive car rental service. Macau Mokes have been stylised to emphasise the image of a fun holiday, they are brightly coloured, fun to drive, and they really set the theme for your Macau holiday. It's not so much a question of car rental to get you from A to B — but more a means of having fun along the way!

Macau Mokes take pride in being more than just a car rental company. Our staff are trained to give assistance and information on where to go and what to see. Each client gets a "MOKEPAK" — our own information package with a guide book, maps, leaflets, etc. In other words, we don't just give you a self-drive Macau Moke — we also give you plenty of ideas on how to enjoy your holiday.

Macau has something for everyone — fascinating history, narrow streets and markets which are great for shopping. For quiet countryside and uncrowded beaches take a trip to Coloane island.

Macau is famous for it's food and wine. The Portuguese explorers of old were great gourmets and the tastes of Brazil, India, Africa and Malaya can all be found in Macau, together with marvellous Chinese food and many many other styles. Execellent wines and ports are freely available everywhere.

Macau Mokes offer a range of hotel/self-drive packages to suit different budgets — and all are excellent value for money. Turn this page over for full details of all our services.

No Far Eastern holiday is complete without a visit to Macau. It's different, it's relaxing, it's fun and it's fascinating!

MACAU MOKES SELF-DRIVE CAR RENTAL WHEN YOU'RE IN MACAU — DON'T GO WITHOUT US WE'RE THE ORIGINALS!

Unusual Mokes

The Mokes bodyshell design easily lent itself to modification, and it was often adapted to very different uses from those originally intended. Many of the accessories in the UK were provided by the Barton Motor Company in Devon, a Morris agent. They not only designed and sold a fibreglass roof and doors for the Moke, but also sold adapted Mokes with electric winches, crop spraying units and auctioneers rostrums amongst others.

The Moke has also been used as a film camera platform, and a few were converted into six wheelers, like the one pictured below, to enlarge the flat loading area. Another more interesting major modification was the short Moke shown below - its turning circle was quite impressive!

The following pages show some of the other activities to which the Moke has been put.

photo by Paul Beard

photo by B.E. Clark

MOKES

1. Out in the paddock something grazed. Jester accepts his steel stablemate without turning a hair
2. Beach Buggy waits for passengers outside the London Hilton
3. You can lead a Moke to the water ... Sidescreens and electric winch from Barton Motor Co. Ltd.
4. In Paris you can hire a Moke on a self-drive-hire hourly basis for seeing the sights
5. Crop spraying is just another job for the Moke ...
6. ... and you can turn it into a mobile lubrication unit
7. Bristol Motor Co. Ltd. market this Mini breakdown truck
8. Going, going, gone! Mobile auctioneer's rostrum, complete with public address system, from Barton

Barton Motor Co. Ltd., Hyde Park Corner, Plymouth, Devon.
Bristol Motor Co. Ltd., Ashton Gate, Bristol.

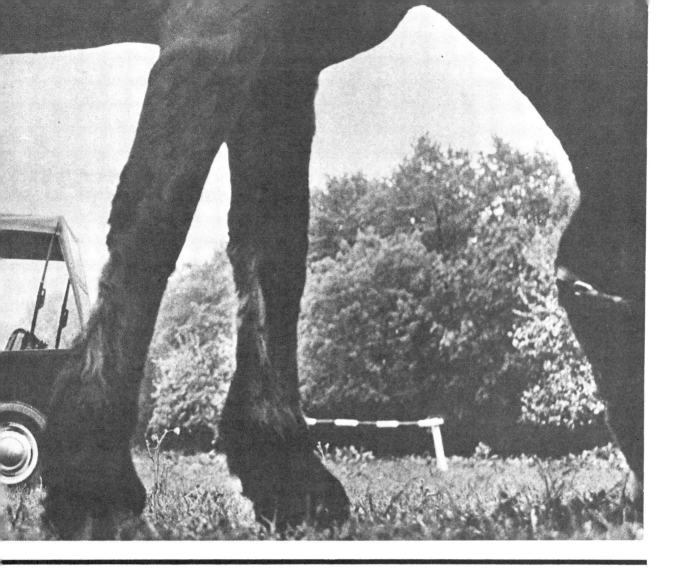

5

6

7

8

Moke Pick-up

Get a load of Leyland Moke Pick-up

Moke Pick-up has a multitude of applications. It's ideal for general farm work. Mining, Surveying. Pick up and delivery. You name it, Moke Pick-up can handle it.

Here's why: the tray top (with the standard drop sides) is 55" wide and 59" long (1448 mm × 1498 mm) and sits on a box-section, all-steel body. It's incredibly strong and rigid.

Like the Moke, its compact length and small turning circle make it ideal for working in tight spaces. It has front wheel drive for superior traction, 8" ground clearance, and an all steel sump guard. And the body is protected by the same electrophoretic rust inhibiting system.

All this plus the same low insurance, low parts prices and general running costs. And up to twice as many miles per gallon as the "average" Australian ute.

AUSTIN mini-MOKE

Barton ACCESSORIES

MORRIS mini-MOKE

THERE ARE NOW OVER 100 ACCESSORIES FOR THE

B.M.C. mini-MOKE

THIS ILLUSTRATED LEAFLET SHOWS SOME EXAMPLES, AND MORE ARE BEING DEVELOPED.

OTHER ACCESSORIES AVAILABLE NOW INCLUDE TWIN WINDSHIELD WIPERS, HAND THROTTLES, LIMITED SLIP DIFFERENTIALS, 5-SPEED GEAR BOXES, MUDFLAPS, RUBBER MATS, ETC.

33.

34.

35.

36.

33. FRONT TOWING BRACKET
34. REAR TOWING BRACKET
35. ELECTRIC WINCH
36.   with 1500 lb. (680 kg.) line pull
37. ELECTRIC WINCH KIT
     with Ground Anchor
38. TERRATIRES
39. CROP SPRAYER
40.   with Boom

37.

38.

39.

40.

1. SIDESCREENS
2.   with Opening Doors
3. WINDSHIELD AIR VENT
4. SPECIAL SEATS
5. HEATER — 3.75 kw.
6.   with Demister
7. FIBREGLAS CAB
8.   with Sliding Windows and Pockets.

9. FIBREGLAS SEDAN
10. FIBREGLAS SEDAN
11.   with Sliding Windows and Pockets
12. FIBREGLAS VAN
13. Bookmakers or Auctioneers' ROSTRUM
14.   with Public Address Equipment
15. BREAKDOWN TRUCK
16.   or WRECKER

17. MINI TRUK
18.   with Half Doors and Screens
19.   with Tonneau Cover
20.   with Drop Tailgate
21.   with rear Hardtop
22.   and Drop Curtain
23.   with Fibreglas Cab
24.   with Fibreglas Cab

25. MINI TRUK – All metal
26. BEACH BUGGY
27. BEACH BUGGY
28. BEACH BUGGY
29. FOAMITE FIRE TENDER
30.   – Dry Powder Type
31. } TECALEMIT TRANSPORTABLE
32. }   LUBRICATION UNIT

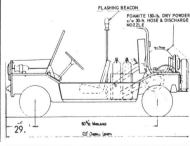

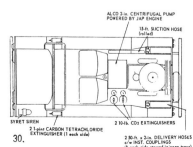

MOKE — DOWN

THE BMC MINI-MOKE IS REALLY AN 'ALL-PURPOSE VEHICLE', AND IN OUR LAST ISSUE WE depicted just a few of the many tasks that it is called upon to do—some ordinary, others not so!

Continuing the same theme, we show on these two pages what our Australian cousins are doing currently with Mini-Mokes. And it certainly makes one think. Is there anything that this 'four-wheeled baby' cannot do?

The picture (top left) shows a Moke in full song during a 12-hour race at Surfer Paradise, Queensland. It surprised all the experts by recording really fast times, thus earning itself a future place in the Australian racing scene.

UNDER

Whilst (left) five Mokes stand in grand array awaiting the officials responsible for running a recent Perth Show. The Mokes were used extensively throughout the Show by the officials in conducting ring events, and a thousand-and-one transport tasks around the showground.

Taffy the Lion (popular television identity with Channel 7, Perth) found that he would have been lost without his Moke (see picture bottom left). This was one put at his disposal by Messrs. Winterbottoms, one of BMC Australia's Distributors.

Taffy glided up and down and around and about during the Royal Show which was also staged at Perth. By the look on his face, Taffy the Lion found the vehicle extremely useful. Needless to say, it created a lot of interest with show visitors and no doubt made many think of buying one—a Moke, not a lion.

The next picture (bottom right page 18) depicts the hunting Moke! Port Darwin Motors supplied this vehicle and it instantly met with success.

The water-buffalo hunters of the Northern Territories have up to now been dogged by lack of transport, for the area is impassable for the majority of vehicles. But the Moke has changed all that, for it will traverse every inch of the country full of equipment and supplies with ease.

Proof of this is that Port Darwin Motors have at the time of writing sold 40 to the residents of the Northern Territories. And that's fact!

The last in this series of 'Moke jobs' comes with the two pictures of the Moke on rails (above). A Mini Mokomotive—this one from Mitchell Motors—was recently put into service by the Tasmanian Railways, and it is, to say the least, one of the most interesting applications we have seen of this multi-purpose vehicle.

The steel flanges are easily fitted and the Moke is then ready for track maintenance duties. The flanges can be just as easily removed for road work.

This adaption has caused a great amount of interest 'down under', so much so that the Railway Authorities are considering running a fleet of Mokomotives.

There you are. A few more interesting tasks that the Moke can deal with. Gives one ideas. Incidentally, we should like to learn of any applications in this country.

MOKE AT LARGE

The two pictures on this page show the Moke in New Caledonia. Left, a cameraman finds it quite simple to lean out and record the trip—just one of the vehicle's many advantages

THE MORRIS MINI-MOKE IS A VERSATILE little vehicle, and looking through a brochure describing it one sees such words as 'fantastic', 'reliable', 'super', 'safe', 'rugged', and 'tough'. Strong words, but words that have a ring of truth about them, for the following stories concerning the Mini-Moke certainly back up the brochure—and these tales are just three of many!

The first tells, mainly in pictures, the story of an expedition through the vast wilderness of the southern tip of New Caledonia. This was a joint venture between two young servicemen and two members on the staff of our Distributors for that area—Société des Automobiles Tracteurs et Matériel Agricole.

The trip took three days over some very rough terrain. Although not dense jungle, or of a really tropical climate, the area covered was an enormous mass of iron-stone. The trip was recorded by one of the servicemen, who in civilian life is a member of Monte Carlo Television, and it can be seen clearly that the Moke had to cover some bad patches —so successfully that another trip of equally tortuous nature is planned for the near future.

The second in our series of Moke tales comes from South Africa. Here the proprietors of the Hotel Langham, one of Johannesburg's leading hotels, have purchased a Moke and put it to a unique use.

The vehicle sports a traditional South African *bakkie* on the rear, and pulls a trailer. It can regularly be seen about the city making house-to-house deliveries for the hotel's off-sales department. It is without doubt an attractive outfit

Still in New Caledonia. This time the Moke is captured negotiating a make-shift wooden bridge. Ahead the ground seems to rise to the very sky, but this stout little vehicle took it all in its stride

From the wilds to civilization. Here a 'domesticated' Moke is found working peacefully in a great South African city. A general runabout—but just another task that shows the versatility of this small 'workhorse'

which serves not only a utilitarian purpose, but also provides a most effective advertising platform both for the hotel and the Moke itself.

The third story stems from home, from the Midlands in actual fact. This is a modern tale but one with ancient connections. The Midland ATV magazine team—ATV Today—decided that a story on the Fosse Way, an old Roman road that runs down through England, would make good material for the programme. One of the team, Lionel Hampden (a well-known TV figure), had the idea that a Mini-Moke would serve them well as a mobile filming platform. This proved to be quite correct as the supporting pictures bear out. So consequently, all along the length of the Roman road, where once trod the feet of legions and where chariots once raced, the wheels of the Moke steadily turned supporting its camera team who recorded a programme which was enjoyed by viewers throughout the Midlands.

There, then, are just three of the Moke's uses, and there are many more.

Truly a versatile little vehicle. Or, as the brochure says, 'A rugged runabout with a thousand uses'.

Cameramen again find use for the Moke. This time a team from ATV Midlands utilize the vehicle as a mobile film platform. The pictures above and left give two different angles

ATV Midlands again, on this occasion they filmed a fox-hunt. The Moke provided their mount and kept pace with the riders across field and dell

Adventures by Moke

Some say that owning a Moke is an adventure in itself. True or not, driving a Moke always brings out a desire to go somewhere a little bit different, be it up mountains, across beaches or through water.

Many outstanding trips were made. In Australia, Hans Tholstrup circumnavigated the country, a trip that involved crossing the Hobart Straits from Tasmania in an inflatable boat containing the Moke. Others, as you will read, decided to cross continents in the vehicle.

In the hands of magazine reviewers it was no exception. Here was a road-going vehicle that they could use beyond the tarmac. For this car it was to be no boring road-test up and down the dual carriageway measuring petrol consumption, acceleration and gearbox sensitivity; what they wanted to know was how far they could go in it and how much they would enjoy the experience.

On the following pages you will see how far the Moke travelled and how true was the Leyland advertising slogan the accompanied the photograph below: "Freedom is priceless, Enjoy it with Mini Moke."

It cruises at 65 mph and, although noise level is high, the ride is pleasant and the hardtop keeps out the weather.

PAT HAYES GOES

HUNTIN' FISHIN' SHOOTIN'

IN A MINI MOKE

Deep creeks such as this one seemed impossible to cross but the Moke proved everybody wrong by going through easily.

The Mini Moke was simply begging to be taken on safari — especially when it came equipped with a hardtop and a heater and had been painted duck-egg blue.

CAR manufacturers dream up some strange names for their brainchilds. Take the Mini Moke, for instance. Everybody knows what "mini" means, be it skirt or motorcar, but "moke", unless you happen to be a reader of Zane Grey Westerns, is a bit trickier. According to the Oxford Dictionary, it means donkey.

So BMC has called its jeep type version of Mr. Issigonis's east-west-engined, rubber-suspended runabout "little donkey". It proved to be one of the most apt titles in the business.

When the Moke was first released in Australia a few months ago it set me thinking in terms of safaris, big game shooting, camping out under the Southern Cross, great white hunters and all the rest that Hollywood has done to death in 10 varieties of wide screen extravaganzas.

When I'd gone home and sobered up a little from BMC's usually lavish hospitality, I reminded myself of leaking canvas side curtains, frostbite, dust, ants and the fact that I'd never even been a boy scout. I didn't possess any means of shooting the big game if I found it in any case.

And that is where this story idea should have flickered briefly and died . . . if I hadn't met Alan Pilkington and been inveigled into drinking (will I ever learn?) with BMC public relations man Mike Quist.

Pilkington, whom I met up with innocently enough at the Melbourne advertising agency where he works, turned out to be a mad-keen hunting, fishing, shooting fanatic. Using subliminal tactics like photographs of monster deer he'd shot and wild stories of even bigger trout he'd caught he managed to get me hooked again on the safari idea.

Pat Hayes, trusty weapon in hand, poses with the Moke outside the log cabin stockmen's hut in Gippsland where the trek finished.

Driven hard enough, the Moke will clear just about any obstacle, even if it gets all its wheels off the ground.

Then Mike Quist came along with news of a weatherproof Mini Moke with real open-and-shut doors, windows, a steel roof and no leaky side curtains. When I learned that its drab green color had been resprayed to an enticing duck-egg blue, and its fittings included a heater, my last vestiges of resistance vanished.

And so Pilkington and I found ourselves battling our way through Melbourne's peak-hour traffic one night in a duck-egg blue Moke loaded to the gunwales with rifles, food, sleeping bags, a tent and all the rest of the paraphernalia for a six-month safari and which we reckoned would last us the four days we'd managed to wheedle from our wives and employers.

That Moke had the same effect on the peak-hour people of Melbourne as a real little donkey would have had. Most grinned, some waved, and smart young things looked as though they'd even pass up a ride in an Alfa for a quick jaunt around the block in it.

Even Melbourne's gruff cab drivers waved us through when we approached under their port bow; a burly truck driver leaned out of his cabin and patted it on the bonnet like a puppy and a little girl tried to feed it a biscuit (True!) until her mother dragged her away.

After cheering up the city dwellers we settled down to the grim task of driving 150 miles or so into the wilds of East Gippsland and Victoria's virtually uninhabited mountain country.

It was admittedly difficult for us to fit ourselves into the roles of pioneers, as the Mini's rubber suspension gave the Moke a reasonably soft ride on highways. With the hardtop conversion keeping out the breeze and the fan-

MINI MOKE
SPECIFICATIONS

ENGINE: Four cylinders, overhead valves. Bore and Stroke: 64.58 mm. x 76.2 mm. Capacity: 998 cc. 60.96 cu. ins. Compression ratio: 8.3 to 1. Carburetion: SU semi-downdraught. Output: 38 bhp at 5250 rpm. 52 lb. ft. of torque at 2700 rpm.

TRANSMISSION: Four speed gearbox with synchromesh on top three ratios.

SUSPENSION: All independent. Front—Rubber springs and telescopic shock absorbers. Rear — trailing arms, rubber springs and telescopic shock absorbers.

BRAKES: Hydraulic drum.

STEERING: Rack and pinion. Turning cricle: 30 ft.

TYRES: 5.20 x 10.

FUEL TANK: $6\frac{1}{4}$ gallons.

CHASSIS: Integral.

DIMENSIONS: Wheelbase, 6 ft. 8 ins.; Track, front — 3 ft. $11\frac{3}{4}$ ins. Rear — 3 ft. $10\frac{7}{8}$ ins.; Length, 10 ft. $6\frac{1}{2}$ ins.; Width, 4 ft. $3\frac{1}{2}$ ins.; Height, 4 ft. 10 ins. Ground clearance, $6\frac{1}{8}$ ins.; Weight, 1867 lbs.

Considering its price the Moke is the ideal kind of vehicle for negotiating this kind of country.

The hardtop conversion's doors can be locked open to provide plenty of fresh air and access to the interior.

HUNTIN', FISHIN', SHOOTIN'

driven recirculatory heater keeping out the cold, it was a fairly comfortable trip.

Despite its low final drive ratio, the Moke cruised quite happily at 65 mph and we found the high, pannier-box sides of the body ideal for mounting a large pile of steaming fish and chips for nourishment during the trip.

Noise level was high, but since the Moke's square-framed constructional metal hardtop didn't have a skerrick of sound-proofing, and the tyres were Dunlop's knobbly winter treads, this wasn't surprising.

Fuel consumption along the highway worked out at 28 mpg and this, too, was reasonable considering our high cruising speed. The interesting point was that it didn't drop much below this, even in the really rough stuff.

After spending a night freezing in our sleeping bags on camp stretchers in a picnic shelter — our pioneer ancestors must have been pretty hardy souls, I've decided — we linked up with a Land Rover manned by two sturdy country men with a chain saw and an axe to complete our journey along a seldom used jeep track to the Moroka River.

An unusually heavy snowstorm a few weeks before had blown trees down all along the track and the small creeks formed by the melting snow hadn't done the road, or what there was of it, any good at all.

Goaded on by the grins of the characters on the Land Rover we charged the Moke at precipices of loose rock and soft mud and, as much to our surprise as to theirs, climbed them with little difficulty.

But worse was to come. A few hundred yards from the stockmen's refuge hut — a small log cabin we'd borrowed for our stay — we were faced by 30ft. of water which was about 30in. deep at its worst. The Land Rover snorted into four-wheel-drive and charged through like a thoroughbred — then everybody looked at the Mini and laughed and laughed.

The Moke was stranded in this creek for some time when the sump guard rode up on to a rock, lifting the front wheels clear.

I climbed behind the wheel and, after asking, without success, for an intrepid adventurer to accompany me, set out into the water feeling like a spaceman heading for the moon on a renovated World War II buzz bomb. As the water became deeper, the soles of my feet were cooled by two jets of water shooting skywards from bolt holes in the floor and I saw the spectators' grins grow wider and wider and then slip into a lopsided kind of amazement as the front wheels churned away at the water and dragged the Moke up on to dry land at the other side.

After crossing the creek, everybody had much more respect for the Moke and, when the Land Rover thundered off into the dusk, Pilkington and I felt more secure about our transport for the next few days.

On our forays around the forests and creeks of Victoria's "high country" we soon learnt that, although the Moke looks like a Jeep, it is NOT a four-wheel-drive vehicle—something, I must admit, that the BMC boys told us before we set out. We were stranded twice.

The first time was when we tried to cross a small creek filled with huge boulders. Inching forward slowly, I managed to work the plate steel sump guard up on to a rock and suspend the front (and driving, remember) wheels in mid-air. We stayed there for some hours until we worked out a solution.

Although two men can lift the rear of a Moke clear of the ground, the front, weighted with engine and gearbox, was too heavy for us. We finally dragged out bits and pieces of tent ropes and tied them all together to make a simple form of "Spanish windlass".

To do this you attach a rope to the vehicle, run it round a nearby tree and back to the vehicle again. Twisting the rope with a stick between the two lengths enabled us to pull the Moke back an inch before we had to untwist the rope, pull it taut around the tree and start again.

It worked, but a small winch would have been easier and quicker.

We became bogged a second time in a deep and particularly soft-bottomed water crossing, but only because we were going too slowly. I soon learned that the Moke will clear practically any obstacle if it's charged with sufficient speed. The jumping and bouncing that ensues does not seem to hurt it and the usually vulnerable sump, gearbox and drive mechanism is well protected by the standard sump guard.

As far as the huntin' and shootin' was concerned, the fallen trees which littered an area of about 50 square miles after the storm gave too much cover to the dingoes, foxes and deer which were in the area. The fishing, however, was much better, as the streams abounded with trout.

I didn't take a rod with me, but managed to get myself a fish for breakfast with a fallen branch and about 8ft. of line which I found in the hut. I attached a hook and worm to the end of the line, dangled it in a creek about 25 yards from the hut and immediately pulled out a respectable fish. I don't know how all the trout fishermen I meet manage to use all their expensive equipment. I suppose they get more fun trying to catch trout in a complicated way rather than using my simple manner.

The hardtop, manufactured by Larsen and Boucher—Transport Engineering Pty. Ltd., is approved by BMC and can be ordered through dealers. It costs $225. It has been cunningly designed to take full advantage of the Mini Moke's unique features.

The wide side doors can be clipped on to the side of the body and the rear door hinges up on to the roof to give plenty of access. Shooters will find they have a wide range of field to aim in without leaving the passenger seat.

The hardtop effectively kept out rain and draughts, and only admitted fine dust when travelling at speed over really dusty roads. It is sturdily built and stood up well to the harder than normal use it received.

We managed to break a couple of retaining bolts on rough roads when a certain amount of flexing put too much stress on rear mounting points. I think this could be solved simply by setting retaining bolts in rubber.

Once mounted, it would not be advisable to remove the hardtop, but access is so good that this should not be necessary.

With the hardtop fitted, the total cost of the Moke would be $1520, and for that kind of money it's the cheapest rough-going runabout you could buy.

FROM SEA TO SHINING SEA BY MINI MOKE

The Good Lord looks after babies and fools

BY DENNIS SIMANAITIS

PARTWAY ACROSS THE vastness of Delaware, I realized that I had made a dreadful mistake. I mean, I used to own an Austin Mini Moke on St. Thomas in the Caribbean; I drove it all over. So why not buy this really nice 1967 Moke in New Jersey and drive it home to California? In November. Across Delaware. And other places I had flown over countless times in only a few hours. Why not indeed?

Mokes and I go way back, even to the dawn of my writing in these good gray pages (see "After the New Gets Put On Again and Wears Off For Good," R&T, August 1973). Back then, I described the Moke as a runt offspring of Jeep out of Mini. It's easy to see it has no doors or windows and the most rudimentary of tops, but photos don't do justice to its 120.0-in. overall length or 51.5-in. width (more than 28 in. shorter and 14 in. narrower than a Honda CRX, for example). I have since refined my definition to say it looks like a military vehicle for a Munchkin Army.

It happens that Larry Holloway is no Munchkin, though he does devote a considerable portion of his auto repair/restoration/preservation talents to Mokes, and it is from him that I bought mine. In fact, keep 1120 E. County Line Rd., Lakewood, N.J. 08701, (908) 370-3916 handy, just in case you get carried away as I did. Larry tells me he is prepared to provide others with Mokes, ranging from do-it-yourself shipments of rusty parts to really nice examples worth $7000, say.

Larry was my Riding Mechanician, as it used to be termed, from New Jersey to as far as Washington, D.C. We began by dipping a wheel ceremonially into the Atlantic and spent much of the day at Englishtown Raceway where we taped what turned out to be 9 seconds on R&T's "Ten Best Cars" (December 1990 issue) for the *Preview* TV show. Moke served as the camera car for part of it. Finally, we plunged southward into the darkness of New Jersey's Pine Barrens. No Devils were sighted, though the presence of Moke's oversize windshield was soon forgotten and we felt drawn mystically through the night.

Early the next morning, the good folks of the Cape May-Lewes Ferry put Moke right up front on the *M.V. New Jersey*. This gave us a head start across that vastness of Delaware, a corner of Maryland and into our nation's capital.

The reason for said visit was a wonderful exhibit at Explorers Hall of the National Geographic Society. "Automobilia—Fact, Fun & Fantasy" included toys, games, objets d'art, jewelry, books, posters, clothing and other artifacts from the automobile's earliest days right through the present. National Geographic staffers enhanced my own personal fun & fantasy by allowing Moke to be photographed in their courtyard. However, another proposed happy snap at the Jefferson Memorial was scrubbed when a careless cabbie rear-ended Moke; damage was more than superficial but not catastrophic.

It was sunny the next morning and a crisp 42 degrees Fahrenheit when Moke and I left D.C. on our way to Roanoke. Our route was west to Virginia's Skyline Drive and south onto the Blue Ridge Parkway, aptly named as the gentle Appalachians were as pure blue as the sky, only several hues deeper. This proved a perfect Moke road, what with its twisties, 35-mph limit and lack of traffic that allowed me to zip along handsomely without taxing the machinery. But was it ever cold! I didn't believe in the concept of wind chill before, but I do now.

I arrived in Roanoke with a couple of fingernails split to the quick and bleeding from the cold. Worse, Moke had begun a slight thud-thud-thudding on left turns, an ominous U-joint sound, accompanied by a grabbiness on startup suggesting accelerated wear of the top motor mount. To cap things off, when getting out for a provisioning errand, I managed to perform a dime-size skin graft off my forehead onto Moke's leftside mirror. California never looked so far away.

PHOTO BY BEVERLY S. NARKIEWICZ

Thornton Gap, Virginia onto the Blue Ridge; sunny and 45 degrees.

It was then, like in one of those Cary Grant movies, that the angel appeared. He said, "Hi, I like your car."

"Uh, thanks. I seem to be bleeding here or I'd shake your hand." Then I told him about New Jersey, California, R&T, the D.C. cabbie, my fingernails, my forehead and Moke's ominous sounds.

"Gee, *Road & Track*, that's my favorite magazine! Come on up to the house, it's only a mile from here. I'm restoring two Minis, a 1959 and a Cooper, and I probably have some spares you could take with you."

And so it was that Buzz Keene showed me how to check the severity of Mini drivetrain problems, while providing me with a spare U-joint and convincing me that Moke probably wouldn't need one for many more miles.

I left Roanoke the next morning happy of heart, convinced that if I looked for his house again, it would be gone and neighbors would say, "No, that lot's been vacant since the Keene boy saved everyone from a fire back in '32 . . ."

From Sparta to Asheville, North Carolina, I encountered the worst weather of the trip, an ice storm that had me scraping off the back side of the windshield while Moke's single wiper worked valiantly on a tiny sector of the front side. We did a pretty good job as we were the last to get over the Asheville Grade before they

Modest lodging for Moke and me? No. It's Asheville, North Carolina's grand Biltmore House.

closed it. I celebrated at Stanley's Barbecue and Clog-Dancing Bar while Moke rested up under its tarp.

Weather improved the next day, and before leaving Asheville, we got to visit some secondhand bookshops as well as Biltmore House, the largest private residence in the U.S., built in 1890–1895 for George Washington Vanderbilt and now a National Historic Landmark. Just the place for a happy snap; you can see Moke if you look really carefully.

A portion of the run to Chattanooga was along the Ocoee River through the Cherokee National Forest, beautiful with autumn colors at their peak. That night, I bedded down in an elegantly vintage parlor car at the Chattanooga Choo Choo Holiday Inn, a finely restored railway station and another Historic Landmark.

Moke and I picked up the Natchez Trace in south-central Tennessee and headed down toward Tupelo, Mississippi, birthplace of Elvis Presley. My firstborn has a hobby of collecting tacky Elvisiana, but Suz would have been generally out of luck here. I found Tupelo and its environs to be tasteful indeed, particularly after standards set in the Great Smokies.

More than 400 miles long, the Natchez Trace has been the route of choice for buffalo, Choctaw, Spanish Conquistadors, French explorers and various riffraff of our own kind. In particular, boatmen would float goods down the Mississippi to New Orleans, work their way back upriver as far as Natchez, sell off the boat as lumber and trek north up the Trace; the Devil's Backbone, it was nicknamed.

By contrast, I found it a kind and gentle way down to Natchez, itself a kind and gentle town on a bluff overlooking the river. The Briars was a lovely bed-and-breakfast there. The Cock of the Walk, down "under the hill" along the river, had great catfish, mustard greens, pickled onions, hush puppies and beer, as well as a sweet resident cat, Momma, who cadged tidbits with the gentlest of meows. And could anyone ever tire of watching boat traffic along the Mississippi?

Accept no substitute! In the foreground above, Moke; to the rear, cotton bale. Also seen in our attempt to get across Texas was the World's Largest Roadrunner, Fort Stockton's Paisano Pete.

Contrary to the opinions of some, there's a lot more west of this river than east of it. Most of Louisiana, for instance. Natchitoches was especially picturesque with its cobblestone main street, *Steel Magnolias* having been filmed here. And, in time, Moke and I entered the great state of Texas, a state we would continue to be in for almost four days.

Our celebration in reaching Dallas was multidimensional. First, my wife and pal, Dottie Clendenin, joined Moke and me for the rest of the trip. Second, we stayed at the Omni Melrose, an elegant hotel restored to Historic Landmark stature. Third, we visited Dallas kinfolk (nephew Davey Bob; his wife, Beverly Bob; and their spaniel, Sadie Bob). And, last, Moke got a well-earned pitstop at Argyle Imports in Fort Worth. Argyle's John Littlejohn has a Moke of his own. He renewed our Moke's U-joint, motor mount, oil and grease, and his dog Lucky even checked out an accustomed Moke seat before we headed west.

Once again, good folks sent us on our way with light hearts. A cash register receipt from a gas stop near there put it aptly; it read, "No Long Waits, No Short Weights, Have a Nice Day."

We had been cautioned about the desolation of west Texas, apparently by those who'd never seen real desert. The land around San Angelo (home of a vast Goodyear test facility) and Fort Stockton (ditto Firestone, not to say Paisano Pete, the World's Largest Roadrunner) looked reasonably fertile. In fact, it was along here somewhere that we discovered the genesis of Moke styling, in cotton country, as I recall.

Some 90 percent of our trip was along secondary roads, though there were parts of west Texas and New

We set a Moke record at Bob Bondurant's (and we sure hope no other Moke ever shows up).

Mexico where we had no choice but Interstate 10. Here, we found the big rigs most cooperative, moving way left as they passed. We felt a little self-conscious about it all until, out in the middle of positive nowhere, we passed a fellow carrying a large wooden cross, the other end of which was on a little 2-wheel cart.

Route 70 in eastern Arizona was the best scenery yet. We stayed in the mining community of Globe, a real one in its work clothes. Then on to Phoenix, where we met our friends Jonathan and Beverly Thompson for lunch at the Arizona Biltmore, a place fascinating in its Frank Lloyd Wright influence and the second Biltmore of the trip. Jon and Beverly had driven east to convoy with us the rest of the way home.

Before leaving Phoenix, we visited the Bob Bondurant School of High Performance Driving in its impressive new home at Firebird raceway. The school's Jerry Jensen showed Moke and me the line around Bob's new road course, and, later, we set a Moke record for the place, easy to do as no other Moke had ever been anywhere near it.

That night, we stayed at the Gila Bend Space Age Lodge, which proved startlingly post-Sputnik in its decor. We all piled into Moke for dinner at El Taco Kid, a Mexican truck stop down the road, where we had the best *gorditas* this side of San Luis Potosi.

On to Yuma, where we enjoyed its Main Street historic district; the Theatre in wonderful art deco style; Bandana Books, an excellent secondhand bookshop; and, around the corner, the Garden Restaurant. Before heading west, we stopped in at the Territorial Prison, a harsh-looking place but actually a model of humane administration at the time. Its history, from 1876 to 1909, is replete with great stories, like the time a prisoner almost managed to escape using a saw hidden in his cork leg. An interesting artifact in the museum is a saloon wall clock with its operation and numerals reversed; this, so wary customers could keep an eye on the time as well as on those rascals behind them.

Yuma is right at the Colorado crossing into California, and the view westward was especially moving. My tough little car was going to make it!

Our California route took us through Dottie's native El Centro, to Thermal for date shakes at Valerie Jean's, past Cabazon's huge concrete dinosaurs and, finally, to cupping a hand in the Pacific and anointing Moke's tire as it sat on Newport Pier, not far from our office and 3904 miles from the other shining sea.

In retrospect, I shouldn't have worried about myself or Moke. Early on, two lovely silver-haired ladies staffing the Visitor Center of the Blue Ridge Parkway had quizzed me about the trip.

"Young man," asked one of them, "does that car of yours have a heater?"

"Uh, no, ma'am."

"And it has no more top than that?" said the other.

"That's right, ma'am."

"What will you do if it rains?"

"Well, ma'am, I guess I'll just have to trust in the Lord."

"Yes," said the one brightly, "the Good Lord looks after babies and fools."

Above, souvenir refrigerator magnets keep Lucas Prince of Darkness firmly lodged in voltage regulator residing directly on the other side of Moke's dashboard/firewall. At right, Newport Pier, the balmy evening of arrival home.

Mini-Moke To Wild Wales

WE LOOKED up "moke" in the dictionary. It didn't give it, which just goes to show what a rotten dictionary we've got. So we turned to our other well-thumbed tome, Roget, and *he* gave it alright: it is apparently the same as an ass, donkey, cuddy, hinny, steed, nag, palfrey, galloway, charger, destrier, courser, hunter, barb, jade, hack, pad, cob—well, you know how old Roget does go on. It all seems to boil down to a beast of burden, which is really quite right after all.

So much for definitions. Obviously Roget hadn't come across the British Motor Corporation, because what they call a moke isn't any of these things. It's a sort of motoring-reduced-to-essentials. In other words, you get four wheels, a chassis, an engine to push it all along, and a seat to sit on. That's a moke—the engine derives from a Mini and so it becomes a Mini-Moke, which is what B.M.C. recommend for the odd bit of cross-country motoring. We borrowed one for 350 miles or so of this sort of thing which had to be carried out in wildest Wales a little while ago, because the idea of Moking has been appealing to us for some time.

You readjust your sights a bit when the Moke arrives. For a start, it's rather like World War One flying—there is no way of keeping either warm or dry except by virtue of the clothes you put on. So you put on a kit which makes you look like a cross between an intrepid aviator of the old school and a motor-cycling fly-fisherman, climb up onto the driver's perch and it's hey for the open road, and all that. Within the first few miles the fuel gauge went AWOL. This didn't matter because, easily accessible from the driver's seat, is the fuel filler. This you take off, and the hole it covers up is not just big enough to look down—you can practically climb in and have a sniff about if all else fails.

Both looking and feeling like Nanook of the North without his kayak, we hied ourselves for Wales. It began to rain. The Moke has a hood, but it doesn't much matter whether you put it up or not. The thing is made to be manoeuvred by agricultural persons and so it closely resembles the cheaper sort of tractor when it comes to the creature comforts. With the hood up, no rain gets inside unless it is horizontal-type driving rain, or unless the Moke is moving. With the hood down, no rain gets inside unless (see "A" above) or the Moke is not moving. Dressed for pot-holing as we were, we decided to carry on.

The next thing that strikes you is that every time you're passed, the wipers fail to cope. This is less a matter for mechanical attention than an expression of the simple fact that the spray from the chap that has just past has put as much water on the inside of the screen as there is on the outside from natural causes. Memo—for long journeys in poor weather, wear absorbent gloves.

The Mini-Moke is very light, so that although, once you get off the road, *really* soft mud stops you like a brick wall, there is no real reason why an average sort of chap in good health shouldn't hump it on to the next firm ground. Its hill-climbing ability is pretty good but if the track is deeply rutted you need to be careful that the thing doesn't ski along on its sump-shield, and also means that as soon as it takes all the weight you will lose all motive power from the front wheels.

O.K.? There is something more exciting than driving the Moke—it's being a passenger in one. There comes a point, on all corners which leave the left-hand side of the car outside, when you realise that there is nothing between you and the ground (whistling past underneath at up to 60 m.p.h.) except God's good fresh air. Which may have excellent remedial qualities, but we never heard it was much good for a broken neck. So you realise why B.M.C. put all those grab-handles all over the place—that is all that will keep you on board. Not, mind you, that the driver feels much more secure. The seat is very high, which means that you bounce along with the steering wheel between your knees and facing the prospect of Something Awful happening to you if you hit a bigger bump than usual.

We haven't mentioned yet, of course, that the Mini-Moke is darned good fun. So long as you don't try and make it behave like a tractor it is reasonably practical, too, for off-the-road motoring—it won't tackle the really rough stuff, lacking the power, traction and ground clearance for the job, which is probably why they aren't too popular with the agricultural community. But they reduce motoring to a basic degree which, for us, has a lot of appeal. You can drive 'em in gumboots—in fact, if the weather is at all chancy this is a pretty good idea anyway—and if the interior gets too stinking with mud and muck you simply take off the seat cushions and turn the hose on. Not quite the thing for a theatre visit, perhaps, but we'd love to be able to think of an excuse for having one.

One of the best excuses, natch, is that it handles so well. It goes round corners like they weren't there—and if this is really because it won't go fast enough to get into trouble, that doesn't alter the case. Even in the pouring rain, when we went round corners too quickly because (a) we were too wet to care and (b) because we couldn't see 'em anyway—but we never even came near losing it because it never tried to get away.

We test BMC's Big Wheel Moke in the mountainous backblocks of New Guinea. Story, photos: David Frith

a moke at 8000 feet!

SO there we were at 8000ft. in this Mini Moke — no, no, on the ground, silly, driving through this mountain pass in New Guinea — and driving through terrifically heavy rain, when up comes this fellow in a Toyota Landcruiser, heading the other way through the mud and slush.

"You're not going through in THAT, are you?" he asked. "Yep," said Bruce. "Don't try it," the man said. "There's a landslide section round the next corner that's just a sea of mud. There's two trucks down to their axles, they've just winched another bloke out, everyone else has been turned back and the road's officially closed. I only just got through in low-ratio four-wheel-drive; **you** haven't a hope."

"We'll be right," said Bruce and slipped in the clutch. And we **were** right. The Moke actually rode OVER that terrible sea of mud, too light to sink in, while the bogged truckies, the road workers and the Landcruiser driver looked on in amazement. "Cheeky bugger," said Bruce. "I've half a mind to go back through it and ask him what he thinks of the Moke now."

Why go to New Guinea to test BMC's new big-wheel version of the Moke? **Modern Motor** had two reasons in picking on this location for our latest long-distance test: We wanted to put the Moke through some really tough country to see whether it lived up to

MOKE is centre of New Guinean attention as we wait in highlands village.

MOKE chugs up steep creek bank in Ramu Valley, with highlands in the background. Stony creekbeds were no obstacle; but pulling power uphill is less than old model.

BMC's claims for it — and it did; and then we wanted to see this Australian vehicle in action in one of its rapidly-growing export territories.

As part of the British Leyland international rationalisation plan, BMC Aust. has now become the world's only producer of the Moke, and are getting solid orders for it in places like Fiji, the Philippines and New Guinea. Much of its success story is due to the Australian-designed big-wheel version, which has turned the Moke from the status of a toy into a serious go-anywhere vehicle, as we were to find on a spectacular succession of stony mountain roads, jungle tracks, river crossings and kunai-covered plains.

Preparations

Modern Motor's route was planned for us by Bruce Chew, sales manager for P.N.G. Motors, the New Guinea BMC distributor, located in Pt. Moresby. A tall, easygoing character, he has spent many years in the Territory and jumped at the chance of operating as our guide. He recommended starting from Lae, on the north coast, and driving 350 miles into the Highlands. Some country, this (see "The Vanishing Frontier", travel section this issue): nearly all of it over 5000ft., with many peaks reaching 14,000-15,000ft. — twice the height of Kosciusko.

Because Lae and Pt. Moresby are split by a continuation of the same central spine of mountains, the Moke had to be shipped round the coast from Moresby to await our arrival. We flew in to Lae a few weeks later, marvelling at the rugged nature of the country and wondering how they ever drove any roads through it.

Our Moke was waiting for us, all gleaming white (once all Mokes were green) and we took the opportunity to check over the differences from the old small-wheel job.

The wheels are the most obvious change. In place of the old 10in. wheels are the sturdier, 13in. jobs, giving the Moke a much more purposeful appearance. In conjunction with raising the rubber cone suspension, they increase ground clearance from around 5in. to a whopping 8in. — certainly good enough to clear most obstacles you're likely to encounter in New Guinea.

Protection of car and occupants goes further than it did, too. Front and rear mud flaps are fitted, the drive shafts are protected by rubber shields, and the sump guard has been redesigned, giving greater engine and radiator protection — good thoughtful moves, all of them.

Unlike the two-seater Australian Mokes, this export version came with four seats. As there were three of us going on the test, we removed one seat — four simple nuts to undo and it lifts straight out — and used the space for luggage stowage. It made a very handy area, comfortably taking two suitcases, large camera bag, overnight bag and tucker box. The inevitable Esky fitted compactly into the space behind the rear seat.

Backing and filling around the Huon Gulf Motel where we spent our first New Guinea night, we soon noticed one disadvantage of the bigger wheels — increased turning circle. The old Moke would whip smartly around in 30ft. or so; the new one would do no better than about 38ft. Still, it's a fairly small price to pay for the other advantages.

Off we go

Some of these advantages we found out as soon as we left Lae one steamy, sunny morning. The old Moke, while a good fun car for churning through a field or across the beach, was an irritating tourer; at anything above 45 mph it would be fairly screaming along, making such a deafening noise that conversation was impossible. Around 50 mph was the practical touring limit. But we soon found the big-wheel Moke was a different kettle of BMC fish and the 40 bhp 998 cc. motor would quickly shove it up to 60-65 mph.

Top gear gives 16 mph per 1000 revs, making the theoretical maximum over 80 mph, though in practice we never had it much over 65 — there were damn few stretches where anything like the maximum was possible, and certainly nowhere we could do an accurate speed check. Suffice to say the Moke now has comfortable touring ability.

We shot out along the Highlands Highway, winding at first through cacao plantations, then dense jungle in which bananas and other tropical fruits grow wild, and finally onto the kunai-covered plains that extend 100 miles from Lae to the mountains' edge at Kassam. The road was dusty and corrugated in spots but mostly a reasonable surface.

The Moke's rubber cone suspension

a moke at 8000 feet!

soaked it up well: it offers the best ride over rough stuff of any cross-country vehicle, with the possible exception of VW's country buggy. Certainly anyone who's been jolted around in Land-Rovers or Nissan Patrols would be pleasantly surprised at how smooth it can be.

Even in the rear seat, the ride was surprisingly good. The rear passenger, in fact, was the most comfortable of all, with loads of legroom and able to sprawl out. The drawback was dust. We were travelling with side-curtains on the rear section only, and the dust tended to creep in and swirl around, trapped in the rear compartment. Weatherproofing remains one of the Moke's biggest bugbears; there are, we hear, some optional fibreglass fully-enclosed hardtops available now, and we're eager to try one.

Seats are still pretty basic but surprisingly comfortable. As in the old Moke, they are PVC and canvas fabric, stretched over a tubular steel frame and laced underneath, but the addition of two wads of foam padding make them just a bit more comfortable. Although the laces tended to slip and had to be retightened several times, they gave good support, rather like a deckchair, and we did not suffer from any aching limbs, despite several long, tiring days' driving.

Tough climb

About 100 miles from Lae the dusty grey road turns and abruptly leaps into the Bismarck Ranges through the Kassam Pass. It climbs from 1000ft. to 5000ft. in the space of a few miles, so you can imagine the steep grades and winding, twisting road.

We soon found ourselves down to second gear and grinding along at 10-15 mph. The bigger wheels and the difference they make to the gearing has certainly improved the vehicle's high-speed touring, but it's done nothing for low-speed pulling power. The old Moke would sit on second or third and just keep chugging away over anything — a display of really solid torque — but the new one can fade disappointingly on hills; we wondered whether an increase to the 1275 cc. Morris 1100S motor, or something similar, might not be an idea to match the bigger wheels. Still, we were never really embarrassed.

The engine grew pretty hot, but did not boil, even when we stopped at the top to admire the breathtaking view back down through the pass to the Markham Valley, a lush and incredibly green plain stretching back for miles, and scene of some of World War II's bitterest fighting.

Once through the Kassam and out onto the 5000ft. plateau of rolling, grassy-green hills, we were able to flash along quite fast again.

Indigenous interest in the Moke was high. We went through a section where

BRIDGES over New Guinea's thousands of fast-flowing mountain streams tend to be rudimentary — or non-existent. This girder job is one of the better ones.

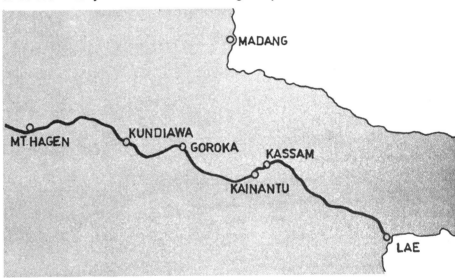

MAP shows the route followed by Moke over New Guinea's Highlands Highway.

the neat little native villages bobbed up astride the road every few miles, and hordes of New Guineans would come streaming down to the road, waving and cheering as though we were Jack Murray going through in the Redex trial. Others would pause in their digging work in the gardens which run up the slopes of the mountains. The kids would cheer lustily; the barebreasted girls would smile and wave shyly; the men would wave solemnly. They are a friendly and happy race, these Highlanders, with handsome looks and faces that light up with tremendous grins.

Petrol worry

We were doing little grinning, however, for we had a fuel supply worry. Goroka, our destination for the day, lay 220 miles from Lae. The fuel tank holds six gallons and the Moke was reputed to do 40-45 mpg, so we hadn't bothered to refuel at Kainantu, the only township on the way, or take cans. In fact the fuel gauge dropped to almost empty when we were still some 30 miles from Goroka, and we slowed down to conserve fuel. We made it, all right, with only a drop left in the tank, which sits inside one of the panniers that form the Moke's sides.

In the circumstances the low-gear work, the mountainous roads and the weight we were carrying (we were three-up, remember), 30-35 mpg was probably a reasonable consumption rate. But a range of just over 200 miles really isn't good enough, and we strongly recommend to BMC that they place a second tank in the other pannier, thereby doubling the range to more than 400 miles. The battery and tools that are now carried in this pannier could be relocated, probably behind the rear seats.

We spent a night in the air-conditioned comfort of Goroka's Bird of Paradise Motel, and headed on early next morning towards Mt. Hagen, 145 miles away.

From here on, the road grew much tougher — narrow, stony and tortuous. Just outside Goroka lay the Daulo Pass, the road itself climbing to 8150ft., while 13,000ft. mountains tower even further over it.

If we thought Kassam was impressive, this was murderous.

MOKE'S new big wheels and increased 8in. clearance have given it true cross-country capabilities, as we discovered on wild variety of New Guinea surfaces. ABOVE: Moke slithers down high-crowned track that would have bothered earlier model. TOP LEFT: Thick mud in notorious 8000ft. high Daulo Pass couldn't stop it either — the lightweight Moke rode over the top. TOP RIGHT: The Moke is dwarfed by 6ft. kunai grass in Ramu Valley. LEFT: Fording fast-flowing mountain stream in the Chimbu country.

MOKE electrics are well waterproofed for crossing streams . . . like this one, forded in Ramu Valley, about 130 miles out of Lae.

a moke at 8000 feet!

PART II

Editor David Frith concludes rugged New Guinea mountain test of BMC's big-wheel Moke

NEW GUINEA'S Daulo Pass is the highest road in Australian territory. When it rains, which is every day or so, it can also be one of the toughest. The road has been carved, mainly by local labour, over—and I mean OVER—an 8000ft. mountain near Goroka in the Eastern Highlands. It's a clay surface, and in places where it runs above a ravine, streams undercut it, causing frequent landslides; a bulldozer is on more or less permanent duty keeping the track clear — and when it has been raining it turns to a long lake of gluggy mud that frequently bogs the trucks that drive over it.

On our first trip through in the Moke — heading up into the mountains — we were fortunate; it was the "dry" season and there'd been no rain for several days. It was still pretty slushy in spots, and we had to wait at the bottom for an hour or so for the pass to open, but there was no trouble. Well-loaded though the Moke was, it cleared it easily, running at low revs and in second gear.

We were to return in very different conditions a day later. This time, however, the mud section was only several hundred yards long. Beyond that lay several miles of tight, winding road to the summit of the mountain—8150ft. up. The grades presented no real problem to the Moke, though a lot of second and first-gear changes were involved—certainly more than in the old model with its smaller wheels.

At the peak we drew up to take some pictures and to toss back a few stubbies of South Pacific Lager, watched by a curious group of indigenes. Native crowds would gather whenever we stopped and poke their heads inside the cab, nodding solemnly. The Moke seemed to interest them more than most vehicles. "Probably wondering whether they might trade in the village Landrover," said our guide and co-driver, Bruce Chew. "Nearly all villages have either a truck or Landrover (the pidgin term for ANY four-wheel-drive vehicle). They're brought co-operatively by the village, which might save for years for it." He told us of cases where salesmen had delivered tractors or vehicles to a village and had to wait hours while the purchase

a moke at 8000 feet!

MUD in the 8000ft. Daulo Pass was no problem for Moke as long as revs were kept down. This shot was taken in the "dry" ... mud was many times worse when Moke returned in tropical downpour.

price was counted out — entirely in old two-shilling pieces.

From the Daulo Pass onwards the countryside became even more striking. The road wound on round mountainside after mountainside, through native villages, past native gardens set out on near-perpendicular slopes, through dense rain forest, over rushing mountain streams and past waterfalls. The New Guineans who waved cheerily from the roadside or the fields were increasingly unsophisticated and more primitive in dress. There were crowds of them walking the road, the women carrying enormous loads of firewood or billum bags (suspended from their foreheads) loaded with bananas, yams, taro or corn, the men striding on ahead with axe, shovel, digging stick . . . or sometimes, bow and arrow. More and more were wearing native dress, too, and an amazing variety of head-dress: ranging from a single flower, to coronets of flowers, feathers, plastic bags . . . anything that was colorful. The trucks that occasionally came the other way were all gaily decorated with fresh-picked tropical flowers or leaves, too.

The further we went the narrower the road became. Any speed above about 20 mph became dangerous, for one blind corner followed another — and three times we had narrow squeaks as a truck came thundering round the other way. Every time there was nothing else to be done but brake, hope for the best, and back up to a suitable parking spot. The New Guineans are pretty skilful drivers, but they go at a suicidal rate.

But as the road became narrower and, incidentally, stonier — it also became more colorful. For the best part of 150 miles to Mt. Hagen it was lined continuously with flower beds — a curious mixture of brilliantiy-colored tropical shrubs and imported varieties ... hollyhocks, cosmos, even roses. The villages all help look after the road, and vie with each other to keep their section beautiful.

We paused briefly in Kundiawa, a spectacularly-sited township in the breathtaking Chimbu region, to photograph some of the mountain people wandering its streets, vividly garbed in native dress — some with bones through their noses and bird-of-paradise feathers on their heads — and plunged on deeper into the mountains.

We had one bad moment when some rudimentary road-making operations blocked the road up a very steep, short hill. A native-owned truck was bogged half-way up and took half-an-hour of shoving to clear. No sooner was it out of the way than three tip-trucks dumped their loads of mud clods on the churned up surface.

It looked impassable, for some of these "clods" were 2ft. in diameter. Bruce, however, just slipped the Moke into second, kept his revs right down, and the Moke slipped and lurched its way up the hill without incident.

The wait at the Daulo Pass and the slow, winding nature of the road had put us some hours behind schedule, and it was almost dusk as we came over a ridge to see before us the magnificent broad stretches of the Wahgi Valley ... a vast expanse of rolling plains, cut into checkerboard patterns by the native gardens. Broad silvery-colored rivers meandered lazily across the plain, disappearing into the purple mountains that bounded it. Smoke from a score of village cooking fires rose slowly into the air to join the

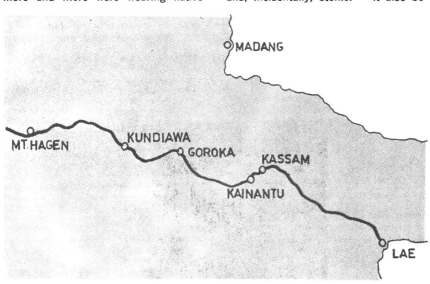

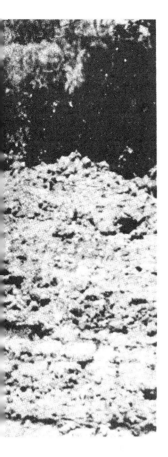

MOKE bowls merrily along little-used track on Ramu Valley cattle property. Occupants appreciated its exceptional ride over really rough surfaces, and nippiness of handling.

mists on the mountains, and when we stopped we could hear native yodelling calls passing from village to village. It was a place and a moment of quiet, impressive beauty.

From here, the road straightened and widened, and we were able to shoot along at a steady 40 mph, past the villagers hurrying home in the gloom, to Mt. Hagen, our terminal point.

Hagen is a bustling, dusty frontier town, filled with a colorful variety of New Guineans in a curious mixture of primitive and European dress. We found ourselves shopping alongside locals wearing only a bunch of banana leaves and a few pieces of strategically-placed string (the leaves are picked fresh every day).

Others had faces covered completely with mud or ochre. And still others wore bank-clerk dress of shorts, shirt and long socks.

The town is surrounded by tea and coffee plantations, many native-owned, and has quite a thriving little industry going in the native artifacts . . . mainly the Mt. Hagen axes, with smoothed stone blades, and wood and grass handles. There is a bird of paradise sanctuary nearby, and we'd like to have stayed longer, but time was running out, so we pressed on to the return run after a night at the local motel.

Quite a run, it turned out to be. We left Mt. Hagen under cloud. By the time we got to Kundiawa, the cloud had burst, almost literally, rain streaming down and turning the stony road into instant slush.

It was still quite warm, but we had to have the side curtains up. I put the stopwatch on the job — it took roughly two minutes for each side, which is probably reasonable. Getting out, however, took nearly twice as long, for the press studs often defy the best efforts

ABOVE: Another Ramu Valley scene — 8in. clearance was plenty for such stony crossings. RIGHT: Typical New Guinea mountain scenery through which Moke test passed — Mt. Helwig (10,000ft.).

of picking fingers, which need to be double-jointed to do the job at all from inside the car.

As the rain grew heavier it began to trickle into the cab from a point where side curtain, windscreen and body all met... exactly the same point where the first Moke **Modern Motor** tested three years ago leaked. You'd think someone might have done something about the design in the meantime.

The water ran back down the side pannier into the rear compartment, where it made things a bit damp for the rear passenger and luggage before finally exiting via the drain holes in the floor.

Wipers, too, we were finding a nuisance — or rather turning them off. Because the centre-mounted switches are just beyond the comfortable reach of a belted-in driver, they have 4in.-long extensions mounted on them.

Fine — except that it makes the switches very vague, and the Moke doesn't have self-parking wipers. Consequently turning the wipers off at the bottom of their sweep becomes a game of guesswork and repeated attempts. Self-parkers would be much more satisfactory.

As things turned out we didn't often have cause to switch the wipers off. The rain was crashing down, the mud was splashing up, and we crept around the tortuous track at a steady 10-15 mph, dreading the thought of meeting anything coming the other way on one of those blind hairpins, with a 1000ft. drop only a few feet to the side. The winter-tread Dunlops were ideal in the conditions and we never shifted on the slippery road surface... I'd have had heart failure if we had.

By the time we reached the Daulo Pass, edging forward round hairpin after hairpin through the steady rain, it was a ghastly mess of churned-up mud. Officially it was closed to traffic, but there was no one manning the topmost gate, so we slid through and approached the mud section with caution.

A Landcruiser came churning up through it towards us, and its driver looked incredulously at the little Moke.

"Don't try it," he advised us. "There's a landslide section round the next corner that's a sea of mud. There's two trucks down to their axles, they've just winched another bloke out, everyone else has been turned back and the road's officially closed. I only just got through in low-ratio four-wheel-drive; **you** haven't a hope."

"We'll be right," said Bruce and slipped in the clutch. And we **were** right. With revs kept right down low to avoid any wheelspin, the Moke just slipped and rolled and slithered over the terrible sea of deep mud, too light to sink in, while the bogged truckies and the Landcruiser driver looked on in amazement.

The Moke finally lurched off the mountain, in the closing-in gloom, edged around the closed gates to the pass, and sped off into Goroka for a cheery night at the local hotel.

Next day, refreshed, revived and with a load of fruit from the local native market, we headed off for Dumpu, a cattle property in the Ramu Valley, where we were to spend several days.

Owner Bruce Jephcott, a former Alice Springs cattle-man now running some 10,000 head of Brahmin-crosses on 19,000 lush Ramu Valley acres, looked on the Moke with tolerant amusement that changed to a grudging respect as we followed his Nissan Patrol tirelessly around the property. The Moke tirelessly forded streams and rivers, belted along tracks, through paddocks and patches of kunai that dwarfed it. Respect, in turn, gave way to admiration when we made a quick return trip into Lae, 130 miles away, over rough roads, to enable Bruce Chew to catch a plane to Port Moresby. The return trip was tipped to take us half a day. We were back in a comfortable five hours.

We spent another day in the lush green beauty of Dumpu, admiring the cattle, looking at a peanut farm, inspecting the wartime plane wrecks that still dot the valley — Dumpu is just below the infamous Shaggy Ridge where the Japanese were turned back in World War II—or just drinking in the mist-shrouded splendor of the mountains that enclose the valley. Mt. Helwig and Mt. Otto (10,000 and 11,000ft. respectively) were just a couple of miles away; in the distance we could see 15,000ft. Mt. Wilhelm, its tip pink-capped with snow each dawn.

Sheared bracket

Then, reluctantly, we left for Lae ...and the plane home. On our final 130-mile run over corrugated roads a sudden clattering from the rear announced that the bracket holding the exhaust pipe in position had sheared through; it was the only thing to break or give us the slightest trouble in our near-1000-mile test that had taken in mountains, rivers, mud, dust, cross-country work and roads both good and bad. Our fuel consumption, an average 32 mpg, was lower than the suppliers, PNG Motors, expected, but struck as reasonable in view of the fact that we were travelling three-up, with a fair amount of baggage, and largely at high altitudes —with the twin problems of low-gear work and thinner atmosphere.

Improvements

We decided we would like to see these improvements in the Moke:
- **An all-synchro gearbox:** in a vehicle that's going to experience a lot of low-gear work, having to double declutch is an unnecessary nuisance.
- **Better waterproofing where side-curtains meet body:** we were surprised no work seems to have been done on the fairly unsatisfactory hood and curtains. Maybe a fibreglass canopy would be the answer.
- **More power:** The old small-wheel Moke couldn't take much more poke without embarrassing wheelspin; the new bigger wheels could certainly stand some more low-speed pulling power.
- **Extra fuel capacity:** A range of under 200 miles isn't enough these days, and the Moke's buckboard construction allows plenty of room for a second tank.
- **Self-parking windscreen wipers.**

Overall we were mighty impressed with the new Moke. At $1420 ($1289 for farmers who can avoid tax) it is the cheapest vehicle on our roads, yet it has now been developed into an amazing little workhorse that will go almost anywhere the big four-wheel-drives will go. Its ride is splendid, its ground clearance problem has been brilliantly solved, and we see a great future for it, both inside Australia and overseas, where BMC Aust. is building a rewarding export trade. ●

Advertising the Moke

A fun car to drive; a fun car to advertise. That seems to have been the maxim over the thirty years of advertising brochures for the Mini Moke.

For the "normal" car, adverts try to concentrate on speed, good looks and the extra options of the vehicle. Not so for the Moke - it has no speed to speak of, the looks have never been described as elegant, and it is rare to find one with electric windows or rear wash-wipe. No, for the Moke such phrases as "Get your Denim seat into ours", "A major breakthrough in air-conditioning", "A draught without the overdraft" and "For the fun of it" were the slogans used to entice potential purchasers.

The early brochures also tried to promote the Mokes versatility. One of the very first slogans was "A rugged run-about with a thousand uses". These uses provided ample photo-opportunities for Mokes performing different tasks from Golf Caddy to Fire Fighter.

The advertising brochures reproduced on the following pages reflect the change in emphasis of the use of the Moke from from it's early military pretentions through to fun vehicle of the nineties. Read through them and see how well the Moke has met the challenge!

The challenge: design and produce an all-purpose vehicle with the Australian man on the land primarily in view ▲ It must be tough and rugged ■ It must be capable of carrying men or materials without fuss ◆ It must be capable of travelling where there are no tracks ▲ It must be economical in use, simple to service ✕ The Mini Moke is BMC's answer ◆ Unlike other attempts at such a design, the Moke is based on a proven and easily obtainable power and transmission unit ▲ It is unquestionably the lowest priced and most economical answer ever presented to the Australian countryman ➤ Rugged? Go anywhere? ◀ Yes! ➤ Meet BMC's Moke and discover how many things it can do for you!

the BMC mini moke
(MILITARY VERSION)

FRONT-WHEEL DRIVE UTILITY VEHICLE

SPECIFICATION

Engine. In line, water-cooled, overhead-valve, four cylinder; three-bearing counterbalanced crankshaft. In unit with clutch, gearbox, and final drive. Installed transversely at front of vehicle.

Bore	2.478 in. (62.9 mm.)	Compression Ratio		7.2 to 1
Stroke	3.00 in. (76.2 mm.)	Maximum B.H.P.		34 at 5,000 r.p.m.
Cubic Capacity	57.82 cu. in. (948 cc.)	Maximum Torque		48 lb. ft. at 2,500 r.p.m.

Fuel System.—Single S.U. carburetter, type HS2; S.U. electrical fuel pump; air cleaner with paper element; petrol tank capacity $6\frac{1}{4}$ gallons (28.41 litres); fuel filters in pump and fuel tank.

Lubrication System. Full pressure to engine bearings, sump forms oil bath for gearbox and final drive; internal gear-type pump driven by camshaft; full-flow oil filter with renewable element; gauze strainer in sump; magnetic sump drain plug; oil capacity, including transmission, 8 pints (4.5 litres) approximately, plus 1 pint (0.57 litre) for filter.

Ignition System. 12-volt coil, and distributor with automatic and vacuum control.

Cooling System. Pressurized radiator with pump, fan, and thermostat, capacity approximately $5\frac{1}{4}$ pints (3 litres).

Transmission. Clutch, $7\frac{1}{8}$ in. (0.18 m.) diameter, hydraulic operation by pendent pedal. Gearbox, four speeds and reverse with synchromesh on second, third and top; in unit with engine and final drive; central floor change-speed lever. Final drive, to front wheels via helical spur gears and open shafts with universal joints; drive casing in unit with engine and gearbox.

Gear Ratios:	Gearbox	Overall	Final Drive	Road Speeds at 1,000 r.p.m.
Reverse	3.628 to 1	13.659 to 1	—	
1st	3.628 to 1	13.659 to 1	—	4.086 m.p.h.
2nd	2.172 to 1	8.178 to 1	—	6.825 m.p.h.
3rd	1.412 to 1	5.316 to 1	—	10.499 m.p.h.
Top	1.000 to 1	3.765 to 1	3.765 to 1 (17/64)	14.824 m.p.h.

Steering. Rack and pinion; $2\frac{1}{4}$ turns lock to lock; two-spoke $15\frac{3}{4}$ in. (0.40 m.) diameter steering wheel. Turning circle 30 ft. (9.14 m.).

Suspension. Front (includes final drive)—Independent with levers of unequal length. Swivel axle mounted on ball joints. Rubber springs and telescopic shock absorbers mounted above top levers. Top levers roller bearing and lower levers rubber mounted at inner end. Fore and aft location by rubber mounted tie-rod. Rear—Independent trailing tubular levers with rubber springs and telescopic shock absorbers. Levers carry stub shaft for hubs which have twin dual-purpose bearings.

Brakes. Foot—All four wheels, hydraulically operated by pendent pedal with leading and trailing shoes all round. 7 in. (0.18 m.) diameter by $1\frac{1}{4}$ in. (0.03 m.) wide front and rear. Hand—Central pull-up lever which operates on rear wheels. In order to achieve positive braking, a pressure limiting valve is introduced between the master cylinder and rear brakes to eliminate rear wheel lock up in emergency application.

Road Wheels. Pressed steel, four stud fixing; 5.20–10, Weather Master tyres, with tubes.

Electrical. 12-volt, 45 amp. hr. capacity battery at 20-hour rate. Double-dipping headlamps with foot-operated dip switch; side lamps in headlamps with separate bulb; combined rear lamps and stop lamps; separate reflectors; rear number-plate illuminating lamp; single horn, with push on instrument panel.

Instruments. Speedometer, including warning lights to show dynamo not charging and head-lamp high beam position. The various switches, including combined ignition/starter switch, are mounted on a panel in the centre of the fascia.

Bodywork. Pressed steel unitary construction, open-type body with vinyl-treated fabric tilt cover supported by folding tilt tubes. Front and rear wings are flat-topped to enable one vehicle to be stowed on top of another for compact transportation such as in aircraft. Fabricated pressed steel sub-frames, detachable from the body provide mounting at front for power pack/front wheel drive assembly and for trailing arm suspension elements at the rear. Four identical seats of pressed steel construction are all detachable, to provide load carrying capacity, and have a limited range of fore and aft adjustment. Hinged bonnet top is also detachable. Windscreen can be folded down or removed completely.

Leading Dimensions and Data:

Wheelbase	$72\frac{1}{2}''$ (1.84 m.)	Pedal, .5g Stop		65 lbs.
Length (overall)	110" (2.79 m.)	Maximum speed		78 m.p.h.
Length (bumpers removed, spare wheel inboard)	100" (2.54 m.)	Standing start $\frac{1}{4}$ mile		$22\frac{1}{2}$ secs.
Width (overall)	54" (1.37 m.)	Range—road		260 miles (418 km.)
Height (screen up)	$51\frac{1}{2}''$ (1.31 m.)	Range—rough country		200 miles (322 km.)
Height (screen removed)	40" (1.02 m.)	Wading depth		13" (0.33 m.)
Wheel revs. per mile	1070	Track—Front		$47\frac{3}{4}''$ (1.21 m.)
Height (tilt up)	57" (1.45 m.)	Track—Rear		$45\frac{7}{8}''$ (1.17 m.)
Stacking height (steering wheel lowered)	$37\frac{1}{2}''$ (0.59 m.)	Normal gross weight		1867 lb. (847 kg.)
B.H.P. per cwt.	2.03	Maximum gross weight (short hauls only)		2100 lb. (953 kg.)

Availability. This vehicle is available to Government Departments for quantity production only. Certain production variations are available to suit Export requirements.

The goods manufactured by The Austin Motor Company Limited are supplied with an express Warranty which excludes all warranties, conditions and liabilities whatsoever implied by Common Law, Statute or otherwise. PRICES—The Company reserves the right to vary the list prices at any time. SPECIFICATION.—The Company reserves the right on the sale of any vehicle to make before delivery without notice, any alteration to or departure from the specification, design or equipment detailed in this publication. Such alterations are likely to occur at any time.

THE AUSTIN MOTOR COMPANY LIMITED
AUSTIN MOTOR EXPORT CORPORATION LIMITED
LONGBRIDGE - BIRMINGHAM - ENGLAND

rugged run·about with a thousand uses

AUSTIN mini-MOKE

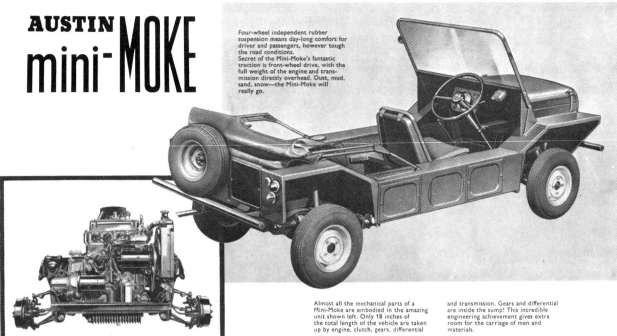

Four-wheel independent rubber suspension means day-long comfort for driver and passengers, however tough the road conditions.
Secret of the Mini-Moke's fantastic traction is front-wheel drive, with the full weight of the engine and transmission directly overhead. Dust, mud, sand, snow—the Mini-Moke will really go.

Almost all the mechanical parts of a Mini-Moke are embodied in the amazing unit shown left. Only 18 inches of the total length of the vehicle are taken up by engine, clutch, gears, differential and transmission. Gears and differential are inside the sump! This incredible engineering achievement gives extra room for the carriage of men and materials.

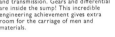

Simplicity of design! That's the hallmark of the Mini-Moke and the reason for its indestructible character. Notice how the weight is where you want it—directly over the driving wheels.

Instrument panel includes warning lights for ignition, low oil pressure, dirty oil filter element and headlights high-beam. Switches mounted centrally.

Just a few of the uses for mini-MOKE

- Hotel beach-wagon
- Holiday camp taxi
- Golf course caddy truck
- Works transport for manager, maintenance men
- V.I.P. factory tours
- Point-to-point transport for works personnel
- Site survey vehicle
- Nimble work-horse for farmers, estate managers, vets.

Never was a vehicle so versatile *and* so economical as the **MINI-MOKE**. Based on the world-proved **MINI Saloon**, it's built rugged all through for sheer reliability.

Cheerfully tackling any task, the **MINI-MOKE** brings the verve of high performance to work-a-day routine. *You* can think of a use for the **MINI-MOKE**.

Fantastic economy, sheer reliability!
—ENGINEERED FOR TOUGH, TAKE-ANYTHING USAGE.

Super-safe handling.
—SURE-FOOTED AS A CAT WHEREVER YOU TAKE IT.

Fantastic economy
Front-wheel pulling power
Superb road holding

sheer reliability!
—ENGINEERED FOR TOUGH TAKE-ANYTHING USAGE

super-safe handling
—SURE FOOTED AS A CAT WHEREVER YOU TAKE IT

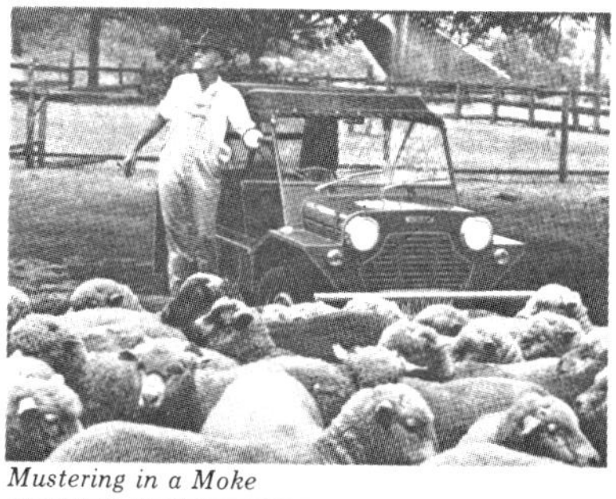

Mustering in a Moke

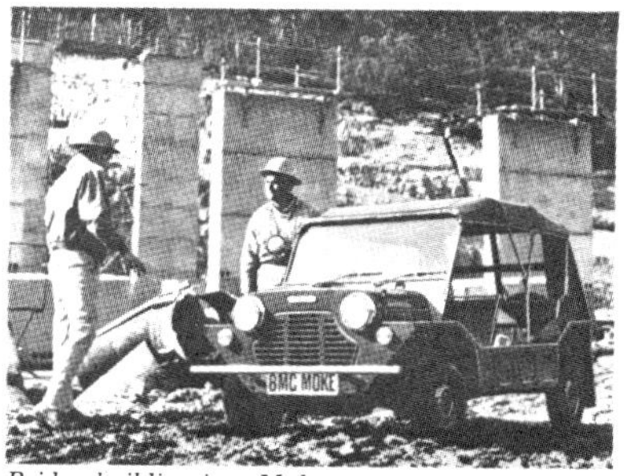

Bridge building in a Moke

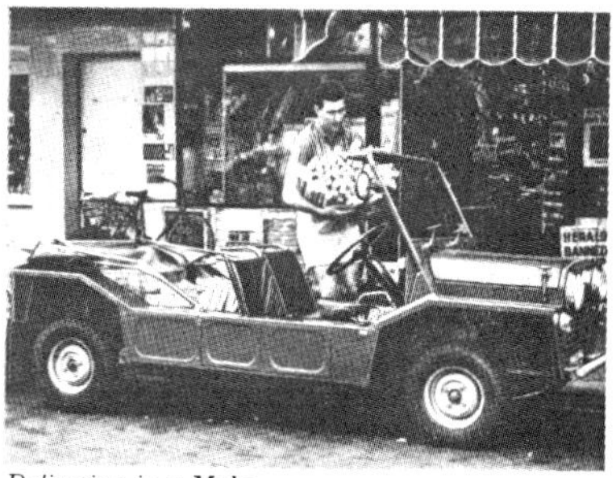

Delivering in a Moke

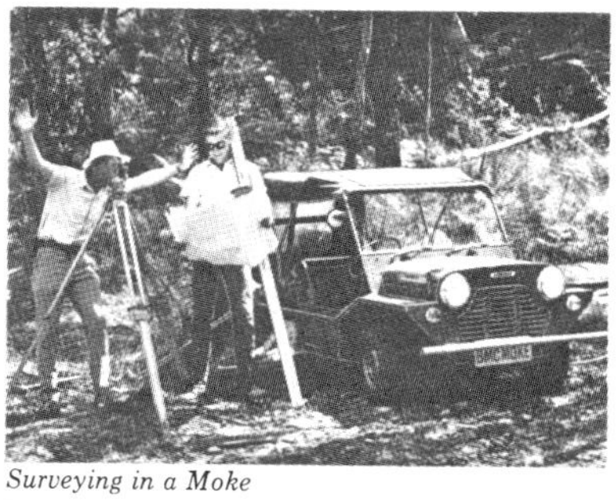

Surveying in a Moke

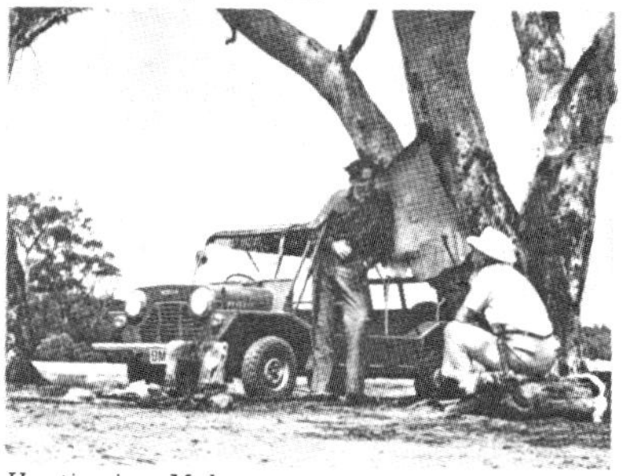

Hunting in a Moke

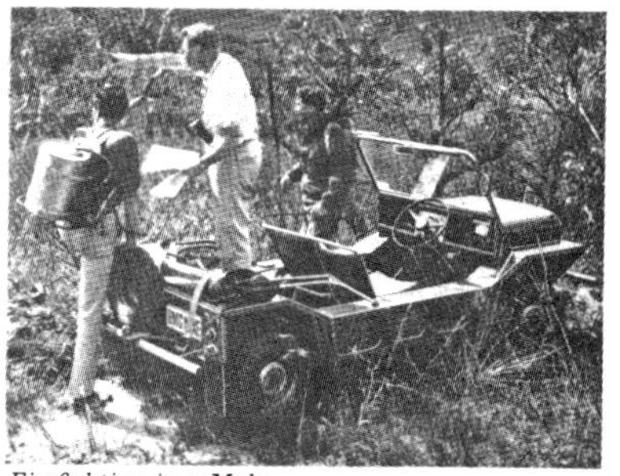

Firefighting in a Moke

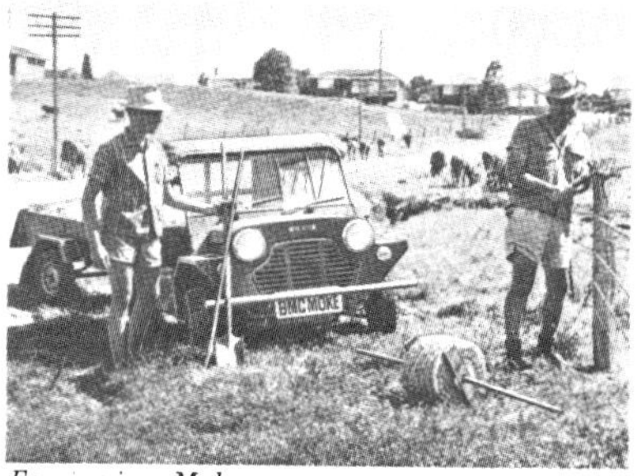

Fencing in a Moke

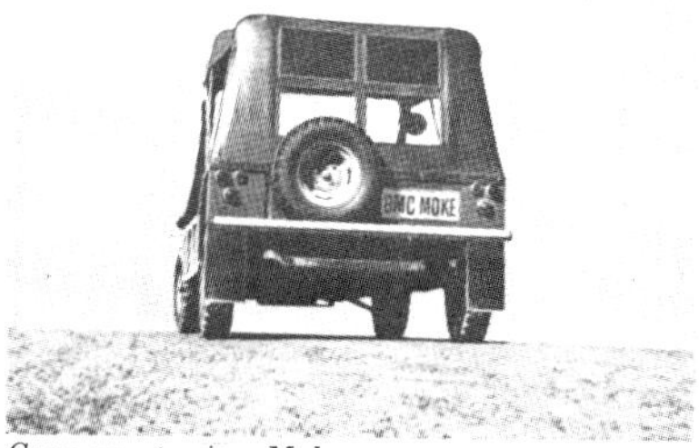

Cross country in a Moke

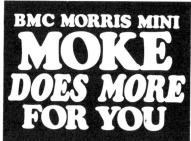

BMC MORRIS MINI MOKE DOES MORE FOR YOU

SPECIFICATIONS

ENGINE: In line, water cooled, overhead valve, four cylinder: three bearing counter balanced crankshaft. In unit with clutch, gearbox and final drive. Installed transversely at front of vehicle.

Bore	2.542"
Stroke	3.000"
Cubic Capacity	998 c.c.
Compression Ratio	8.3 : 1
Maximum B.H.P.	38 @ 5,250 R.P.M.
Maximum Torque, 52 lb. ft. 2,700 R.P.M.	

FUEL SYSTEM: Single S.U. Carburettor, type H.S.2; mechanical fuel pump; air cleaner with paper element; petrol tank capacity 6¼ Imperial gallons; fuel filters in pump and fuel tank.

SUMP PROTECTOR: Sheet steel sump guard attached to sub-frame protecting full length and width of sump.

LUBRICATION SYSTEM: Full pressure to engine bearings, sump forms oil bath for gearbox and final drive; internal gear type pump driven by camshaft; full flow oil filter with renewable element; gauze strainer in sump; magnetic sump drain plug; oil capacity, including transmission, 8 pints approximately, plus 1 pint for filter.

IGNITION SYSTEM: 12 Volt earth return system and distributor with automatic vacuum control.

COOLING SYSTEM: Pressurized radiator with pump, fan and thermostats, capacity approximately 5¼ pints.

TRANSMISSION: Clutch, 7⅛" diameter, hydraulic operation by pendant pedal. Gearbox, four speeds and reverse with synchromesh on second, third and top; in unit with engine and final drive; central floor change-speed lever. Final drive, to front wheels via helical spur gears and open shafts with universal joints; drive casing in unit with engine and gearbox.

GEAR RATIOS:

	Gearbox	Overall	Road Speed Per 1000 R.P.M.
1st	3.627	15.0	3.7 M.P.H.
2nd	2.172	8.8	6.4 M.P.H.
3rd	1.412	5.8	9.7 M.P.H.
4th	1.000	4.133	13.5 M.P.H.
Reverse	3.627	15.0	

Final Drive Ratio: 4.133 : 1

STEERING: Rack and pinion; 2-1/3 turns lock to lock; two spoked 15½" diameter steering wheel. Turning circle 30 ft.

SUSPENSION: Front (includes final drive) — Independent with levers of unequal length. Swivel axle mounted on ball joints. Rubber springs and telescopic shock absorbers mounted above top levers. Top levers roller bearing and lower levers rubber mounted on inner end. Fore and aft location by rubber-mounted tie-rods. Rear — independent trailing tubular levers with rubber springs and telescopic shock absorbers. Levers carry stub shaft for hubs which have twin dual purpose bearings.

BRAKES: Foot — All four wheels hydraulically operated by pendant pedal with two leading shoes at front and single leading shoes at rear. 7" diameter x 1½" width at front and 1¼" width at rear. Hand — Central pull-up lever which operates on rear wheels only. In order to achieve positive braking a pressure limiting valve is introduced between the master cylinder and rear brakes to eliminate rear wheel lock-up in emergency application.

Total Friction Area — 74 sq. ins.

ROAD WHEELS: Pressed steel 3.50B x 10 with four ⅜" diameter wheel studs. 5.20 x 10 Weathermaster tyres with tubes.

ELECTRICAL: 12 volt, 38 AMP. hr. capacity at 20 hour rate. Double-dipping headlamps with foot-operated dip switch; combined rear lamps and stop lamps; separate reflectors; rear number plate illuminating lamp; single horn with push on steering wheel. Combined front flasher and parking lamps with separate bulbs. Separate rear flasher lamps. Dual windscreen wipers.

INSTRUMENTS: Speedometer and fuel gauge. Warning lights show dynamo not charging, headlamp high beam, low oil pressure and oil filter requires renewal. The various switches, including combined ignition/starter switch are mounted on a panel in the centre of the facia.

BODYWORK: Rotodipped, pressed steel unitary construction open-type body with vinyl-treated fabric tilt cover supported by folding tilt tubes. Fabricated pressed steel sub-frames, detachable from the body, provide mounting at front for power pack/front wheel drive assembly and for trailing arm suspension elements at the rear. Two identical tubular seats; made of polyether foam over canvas base covered with P.V.C. coated leather cloth. Load carrying capacity 10 cubic ft.

DIMENSIONS

Width Overall	51½"
Ground Clearance	6"
Wading Depth	13"
B.H.P. Per Cwt.	2.3
Gross Vehicle Weight	1867 lbs.
Kerb Weight	1255 lbs.
A	38"
B	12½"
C	53"
D	58"
E	18"
F	18"
G	17"
H	80"
J	126½"
K	23"
L	23½"
M (Approach angle, front)	38 degrees
N (Departure angle, rear)	46 degrees
P (loading length)	43"-45" depending on seat position
(loading width)	41"
Front track	47¼"
Rear track	46⅞"

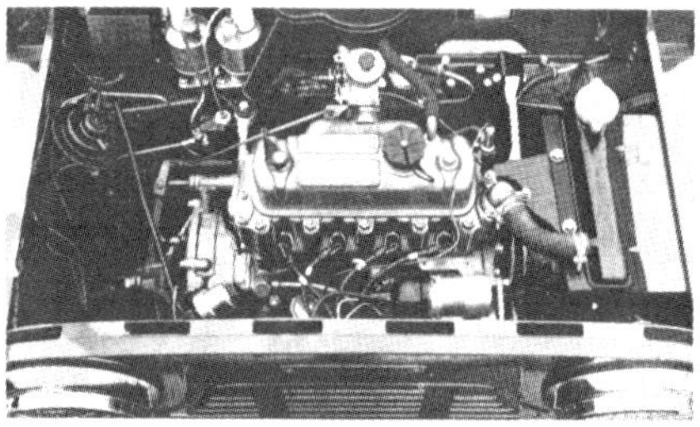

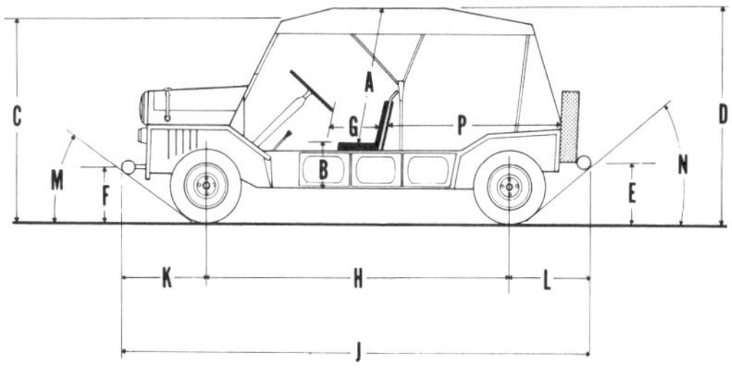

A PRODUCT OF THE BRITISH MOTOR CORPORATION (AUSTRALIA) PTY. LTD.

The goods manufactured by the British Motor Corporation (Australia) Pty. Limited are supplied with an express warranty which excludes all warranties, conditions and liabilities whatsoever implied by Common Law, Statute or otherwise. PRICES: The Company reserves the right to vary the list prices at any time. SPECIFICATION: The Company reserves the right on the sale of any vehicle to make delivery without notice, any alteration to or departure from the specification, design or equipment detailed in this publication. Under present supply difficulties such alterations are likely at any time. The issue of this folder does not contain an offer.

How the Moke meets Australian needs

*Above: The Moke at work on a typical dairy property. Low loading height (22") makes for easy loading of milk cans into back of vehicle. Addition of trailer adds to Moke's versatility.
It takes standard trailer couplings.*

Distance and indifferent roads are part of the Australian motorists' heritage. And nowhere is this more apparent than in the vast open spaces of the outback, in the tough and challenging environment of national development projects. Here there has never been a final answer to transport problems. True, there has been immense improvement in rough-going capabilities of the modern car but it is by no means rugged enough to claim that it can "go anywhere". Of course very special 4 wheel drive vehicles can conquer the toughest going but their cost often makes them an uneconomic proposition.
What is needed is a vehicle that can offer a car's low initial cost and economy of use combined with rugged power to handle rough going. Add a load carrying capacity, ease of getting in and out of the vehicle (every man on the land with gate opening and shutting problems will appreciate this!) and you have . . . THE MOKE!

The Moke is an ideal runabout. Tools, fuel or feed, even an occasional beast may be loaded and carried with ease.
It is economical to operate. It can handle the roughest going with power to spare. It can "duck into town" quickly and comfortably.

In the simplest sense the Moke is a low priced, simple to operate, basic transportation to take over the role hitherto reserved (usually with sad consequences!) for cars or utilities. And in addition its exciting performance opens up an infinite number of uses up till now associated only with vastly more sophisticated and expensive vehicles.

Specialised uses for the Moke

As well as being a general purpose vehicle, the Moke's many features invite its use for specialised applications. These include light delivery work, servicing and almost any job where the basic requirements for vehicle specification are economy, ruggedness and load carrying ability. Just four of such applications are illustrated here.

Top left: The Moke becomes a light delivery and pickup vehicle! Note here the side curtains: a full all weather set of 4 is available as an optional extra.

Bottom left: Specialised applications like line maintenance work are all in a day's work for the Moke. It handles rough going easily because of the flexibility of its rubber suspension and special sump protection.

Top right: Construction projects of all kinds offer endless possibilities for profitable employment of Moke's special talents. What other vehicle offers the economy, convenience and go-anywhere ability of the Moke?

Bottom right: Here the Moke makes an ideal newspaper delivery wagon. There's room for a whole suburb's supply of papers; plenty of room to throw them accurately and easily!

THE NEW BIG WHEEL BMC MOKE
WITH THE NEW 1098cc ENGINE

SPECIFICATIONS

ENGINE: In line, water cooled, overhead valve, four cylinder; three bearing counter balanced crankshaft. In unit with clutch, gearbox and final drive. Installed transversely at front of vehicle.
- Bore .... 2.543"
- Stroke .... 3.296"
- Cubic Capacity .... 1098 cc
- Compression Ratio 8.5:1
- Maximum BHP ...... 50 @ 5,100 RPM
- Maximum Torque .... 60 @ 2,500 RPM

FUEL SYSTEM: Single S.U. Carburettor, type H.S.2; mechanical fuel pump; air cleaner with paper element; petrol tank capacity 6¼ Imperial gallons; fuel filters in pump and fuel tank.

SUMP PROTECTOR: Sheet steel sump guard attached to sub-frame protecting full length and width of sump.

LUBRICATION SYSTEM: Full pressure to engine bearings, sump forms oil bath for gearbox and final drive; internal gear type pump driven by camshaft; full flow oil filter with renewable element; gauze strainer in sump; magnetic sump drain plug; oil capacity, including transmission and filter, 8¼ pints approximately.

IGNITION SYSTEM: 12-Volt earth return system and distributor with automatic centrifugal and vacuum control.

COOLING SYSTEM: Pressurised radiator with pump, fan and thermostat, capacity approximately 5¼ pints.

TRANSMISSION: Clutch, 7¼" diameter, hydraulic operation by pendant pedal. Gearbox, four speeds and reverse with synchromesh on all forward gears; in unit with engine and final drive; central floor change-speed lever. Final drive, to front wheels via helical spur gears and open shafts with universal joints; drive casing in unit with engine and gearbox.

GEAR RATIOS:

	Gearbox	Overall	Road Speed Per 1,000 RPM
1st	3.526	15.056	4.5 MPH
2nd	2.218	9.471	7.2 MPH
3rd	1.433	6.130	11.2 MPH
4th	1.000	4.267	16.0 MPH
Reverse	3.545	15.129	4.5 MPH

Final Drive Ratio: 4.267:1

STEERING: Rack and pinion; 2 1/16" turns* lock to lock; two spoked 15¾" diameter dished steering wheel. Turning circle 36 ft.

SUSPENSION: Front (includes final drive)—Independent with levers of unequal length. Swivel axle mounted on ball joints. Rubber springs and telescopic shock absorbers mounted above top levers. Top levers roller bearing and lower levers rubber mounted on inner end. Fore and aft location by rubber-mounted tie-rods. Rear-independent trailing tubular levers with rubber springs and telescopic shock absorbers. Levers carry stub shaft for hubs which have twin dual-purpose bearings.

BRAKES: Foot — All four wheels hydraulically operated by pendant pedal with two leading shoes at front and single leading shoes at rear. 7" diameter x 1¼" width at front and 1¼" width at rear. Hand—Central pull-up lever which operates on rear wheels only. In order to achieve positive braking a pressure limiting valve is introduced between the master cylinder and rear brakes to eliminate rear wheel lock-up in emergency application.

Total Friction Area—74 sq ins

ROAD WHEELS: Pressed steel 4.50J x 13 with four ⅜" diameter wheel studs. 5.60 x 13 Weathermaster tyres with tubes.

ELECTRICAL: 12 volt, 40 AMP hr. capacity at 20-hour rate. Double-dipping headlamps with foot-operated dip switch; combined rear lamps and stop lamps; separate reflectors; rear number plate illuminating lamp; single horn with push on flasher switch arm. Combined front flasher and parking lamps with separate bulbs. Separate rear flasher lamps. Dual windscreen wipers.

INSTRUMENTS: Speedometer and fuel gauge. Warning lights show alternator not charging, headlamp high beam, low oil pressure, and flasher lamps. The various switches, including combined ignition/starter switch are mounted on a panel in the centre of the facia.

BODYWORK: Rotodipped, pressed steel unitary construction open-type body with vinyl-treated fabric tilt cover supported by folding tilt tubes. Fabricated pressed steel sub-frames, detachable from the body, provide mounting at front for powerpack/front wheel drive assembly and for trailing arm suspension elements at the rear. Two identical tubular seats; made of polythene foam over canvas base covered with P.V.C. coated leather cloth. Safety belts are fitted. Load carrying capacity 10 cu ft.

DIMENSIONS

A	Seat Cushion to Roof	36"	L	Wheel Centre to Bumper Bar (Rear) ... 18¾"
B	Seat Cushion to Floor	11"	M	Breakaway Angle (Front) ... 46°
C	Windscreen Height to Ground (Unladen)	55"	N	Breakaway Angle (Rear) ... 51°
D	Overall Height	60"	P	Loading Length ... 41"
E	Height to Centre of Bumper Bar (Rear)	20½"	R	Loading Width ... 41"
F	Height to Centre of Bumper Bar (Front)	20"	S	Bumper Bar (Rear) to End of Towbar Bracket ... 3"
G	Squab to Steering Wheel	17"	T	Top of Steering Wheel to Ground ... 41½"
H	Wheelbase	83"		Front Track ... 48¼"
J	Overall Length	127½"		Rear Track ... 49¼"
K	Wheel Centre to Bumper Bar (Front)	22½"		Overall Width ... 57"
				Depth of Tray ... 11"
				Ground Clearance ... 8¼"

The goods manufactured by B.M.C. Australia are supplied with an express warranty which excludes all warranties, conditions and liabilities whatsoever implied by Common Law, Statute or otherwise. **PRICES:** The Company reserves the right to vary the list prices at any time. **SPECIFICATION:** The Company reserves the right on the sale of any vehicle to make delivery without notice, any alteration to or departure from the specification, design or equipment detailed in this publication. The issue of this folder does not contain an offer.

Big wheels and an 8 inch ground clearance make the Moke an ideal vehicle for rural work.

Side curtains: A full all weather set of 4 is available as an optional extra.

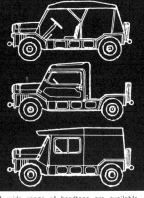

A wide range of hardtops are available, if required, as an optional extra. Your BMC Dealer will supply full details upon request.

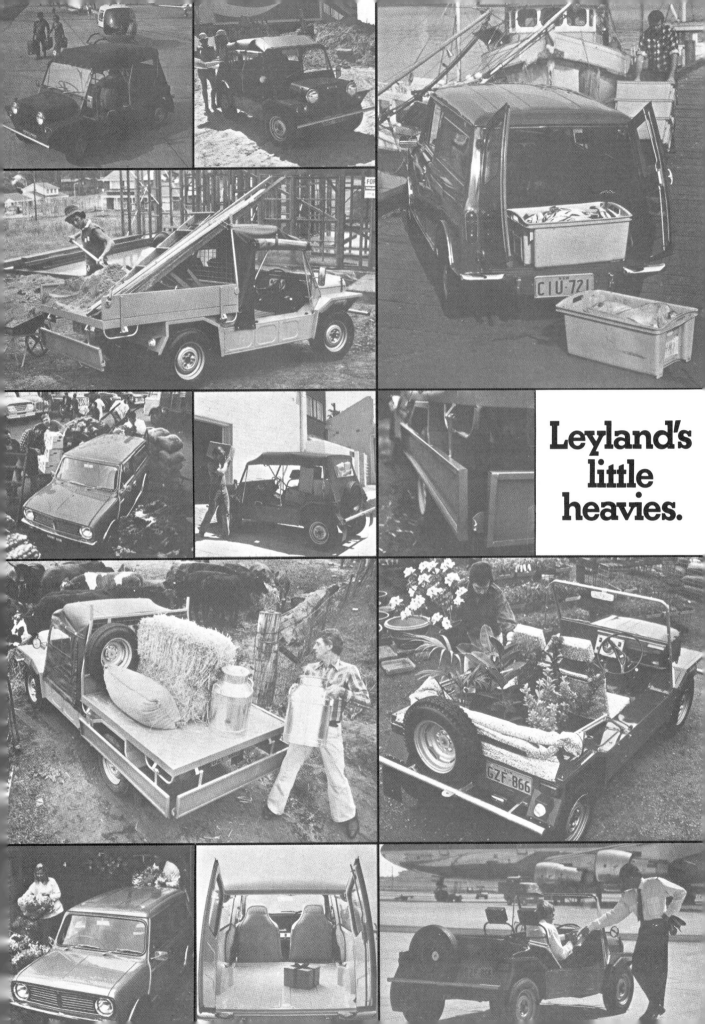

and does it all for next to nothing.

Mini Moke.
It's just about the cheapest four wheels in Australia, and it beats current new ute prices by hundreds of dollars. It's simple to maintain and parts are cheap too.

Mini Moke carries everything you want, just about anywhere you want to go, and gets up to 6.7 litres per 100 km as well.

It's ideal for commercial and recreational use, has a compact size and tight turning circle that makes it incredibly manoeuvrable. Mini Moke is used to carry everything from people to pipeline. You'll find it working on airports, at hotels (carrying people and luggage), on construction sites and, of course, on the land.

With over 200mm (8") ground clearance, rugged, simple box-section all-steel construction, a hefty steel sump-guard and a body protected by a new rust inhibiting system, Mini Moke is built to take the knocks and keep right on going.

If you'd like a bigger engine, take a look at the Moke Californian with the optional 1275cc engine and disc brakes.

Mini Moke Tray Top.
Mini Moke Tray top has a multitude of applications. It's ideal for general farm work. Mining, Surveying, Pick up and delivery. You name it, Mini Moke Tray Top can handle it.

Here's why: the tray top (with the standard drop sides) is 1448mm wide and 1498mm long (55" x 59") and sits on a box-section, all-steel body. It's incredibly strong and rigid.

Like the Mini Moke, its compact length and small turning circle make it ideal for working in tight spaces. It has front wheel drive for superior traction, over 200m (8") ground clearance, and an all-steel sump-guard. And the body is protected by the same rust inhibiting system.

Mini Moke.

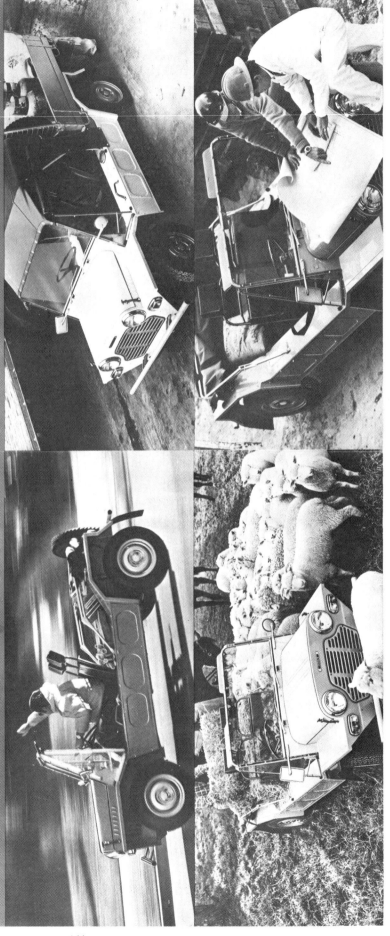

Denim never looked so good. The new Moke Californian, equipped with denim look hood, curtains and deluxe seats and a list of accessories to make it big news on the roads this year.

Bull bars front and rear, a sports steering wheel, front and rear floor mats, spoked wheels, radial winter tread tyres, dual horns, locking spare wheel nut, chrome wheel nuts and centre caps, zip opening rear hood panel, square mirrors paired externally. Metallic paint available as an optional extra. And to top it all – no big bills to spoil your fun in the sun. Moke gives you up to 42 mpg, it's in the cheapest insurance category with most insurance companies and if in your excitement you knock a panel or two, it won't cost a fortune to repair. **Moke Californian. Get your denim seat into ours.**

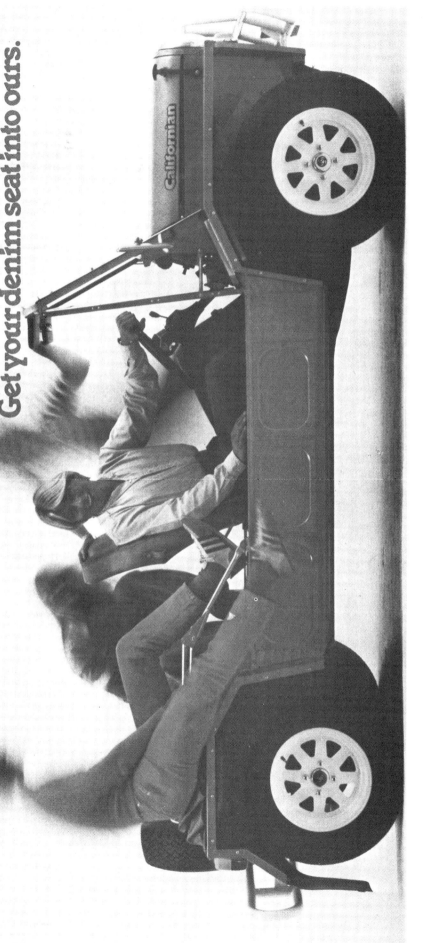

ENTRA NELLA NATURA CON MINI MOKE, più simpatica di un'utilitaria, più agile di un fuoristrada tradizionale. Piccola (solo 3 metri di lunghezza), scattante con il suo motore da 1000 cc; a trazione anteriore, MINI MOKE dà il meglio di sé fuori dalle solite, noiose, strade asfaltate. Procede sicura in riva al mare come sui prati, si muove agile sui terreni scoscesi e sui sentieri impervi, con le sospensioni indipendenti, i freni a disco anteriori, la piastra di protezione della coppa motore. Sempre a suo agio, sempre alla moda, bianca o azzurro metallizzato, con i rivestimenti interni bianchi.
PORTALA FUORI A DIVERTIRSI. Come tutti quelli a cui piace l'allegria, MINI MOKE ama stare insieme agli amici. Ospita quattro persone, due davanti sugli ampi sedili con appoggiatesta, due dietro, sul comodo sedile a panchina. Mentre un baule posteriore, provvisto di chiave, serve a contenere pacchi o attrezzi sportivi.

PERCHÉ MINI MOKE HA UN'ANIMA SPORTIVA. Ma ama anche le comodità: come i doppi specchi retrovisori panoramici e il doppio roll-bar a gabbia che sostiene un morbido top, con discendenti laterali, uguale al rivestimento interno. Così MINI MOKE è sempre pronta: per non farvi perdere nemmeno un raggio di sole, o per ripararvi subito alle prime gocce di pioggia.
MINI MOKE, IL PICCOLO FUORISTRADA PER IL TEMPO LIBERO. BUON DIVERTIMENTO!

The New Generation MOKE

For the fun of it

The cult car of the Sixties is back!
And it's just as much fun now as it was back in those heady days in the Kings Road.
If you remember the original you will find the new generation Moke reassuringly familiar.
It is still nippy to drive - the engine and gearbox are still based on the well proven 'A' Series - it still handles superbly.

It's still easy to park, cheap to run and it still turns heads wherever it goes. However a few things have changed. The wheels are larger and smarter (the SE version features 8-spoke alloy wheels). Disc brakes are now fitted to the front and the engine is far quieter thanks to a redesigned cooling system and electric fan. It is also designed to run on standard lead-free petrol. Front and rear bull bars are essential for any leisure vehicle in the Nineties - so we fit them as standard. There is now a small boot to lock away your valuables.

You and your passengers now have the protection of a sturdy roll over/safety cage.
The use of modern materials means the totally new soft top is more flexible, more durable and easier to fit and remove. You can take the whole top right off, alternatively the sides can be rolled back or removed to give you all the excitement of 'open cockpit' driving while still retaining some protection against sudden showers.

The Moke is a truly 'all purpose' vehicle, equally at home on shopping trips or carrying bales of hay across the paddock.

The fully galvanised monoque body will take all the abuse that off-road adventures throw at it. The Moke is the ideal vehicle to support your outdoor sporting activities.
Whatever sport you are into, be it scuba diving or rock climbing, there is plenty of room for all your equipment behind the front seats.

On the other hand you could just drive the Moke for the fun of it - there's nothing else like it - call us now to arrange a test drive and you will see what we mean.

The well proven 'A Series' power unit showing the uprated cooloing system for greater efficiency and quieter running.

Comprehensive instrumentation with easily identifiable switch gear.

Moke Restorations

When considering a motor car restoration, a Mini Moke has a lot going for it. The combination of flat body panels, minimal upholstery, and the absence of trivialities such as doors and a roof make the task shorter, simpler and cheaper.

The downside is that the pursuit of fun by it's previous owners may not always have been kind to a body which tends to collect water, mud and sand with alarming ease. The prospective restorer may find that there is not be much left that can be preserved. Alternatively, as in a couple of the cases that follow a specialised off-road use may well have removed some of the body work altogether. The restoration can soon become a major rebuild.

Still, the urge to own a Moke overcomes all the problems, and can save you money, as the excellent final article shows. This article, which appeared in Road & Track in 1973 launched the author, (actually Dennis Simanaitis) into a distinguished career in motoring journalism - he is now Road & Tracks engineering editor. He never forgot his Moke and almost 20 years later bought another one and drove it back across the U.S.A., as described in the "From Sea to Shining Sea" article in the Adventures by Moke Chapter.

Pictures by Philip Cooper and Brian Thompson

Making the Moke of it!

Brian Thompson tells the detailed story of his Mini Moke restoration

There are hardly any curves in a Mini Moke's body so DIY panel fabrication is a relatively simple process.

Making the Moke of it!

This engine was taken from a 'donor' Mini. The original was rusted solid.

As an aircraft technician I have a great interest in anything mechanical. When I was a boy if I was not up to my elbows in Meccano, or helping (or hindering!) dad in the garage, I was probably dissecting an alarm clock or performing a post mortem on a portable radio. Still, it was to be the foundation stone of greater mechanical achievements.

I joined the RAF at 17 as an apprentice technician and, in fact, have just finished my engagement after ten years of service. Mini Mokes have long been a passion of mine and wherever I go I keep one eye open for interesting sightings. My service career took me all over the world and, while on Ascension Island during the Falklands conflict, I actually managed to work on a purple Moke belonging to one of the locals!

However, my story began in August 1986 when a neighbour told me about a 1967 Austin Mini Moke that needed a good home. It was being given away and, when I asked about the catch, he said that it was down in London .... in bits! In addition it had to be shifted by the end of the following week or it was to be scrapped.

I hired a car trailer and made the journey from Norfolk to London where the Moke was being stored at an Army Cadets' barracks in Hornsey. The owner had dismantled it but then sadly run out of enthusiasm for the rebuild project. Interestingly, he was the original owner and had used the Moke at

Nobody can accuse Moke drivers of being pampered! Note the trunking for the heater running from the nearside quarter panel.

The Moke as it arrived from London. Not a pretty sight!

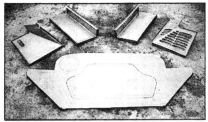

A selection of homemade panels.

North Weald Aerodrome for parascending of all things!

What I found

The side seams were bulging with rust between the spot welds and there were holes in several places. But, even more seriously, all the bodywork forward of the bulkhead had been hacksawed off! There was no front panel, no inner wings, in fact, no engine bay at all! However, I did find a brand new bonnet (which was to prove invaluable), four metal seats, a hood frame, a screen frame (plus cracked glass), five wheels and tyres, both subframes, transmission and a tatty hood. Despite the problems, though, I couldn't resist it so everything was loaded on to the trailer and whisked back to Norfolk.

My next move was to buy a C reg. Mini Minor estate (for £35) for spares. With no front panels I had two options. I either could order them from Runamoke or, alternatively, make them myself. The bought ones would have cost about £300 and, bearing in mind that the Moke cost me nothing, I was reluctant to spend that much on a few panels and plumped for the DIY approach. I was encouraged by the fact that Moke panels have no curves (just straight folds) so I bought a couple of 6ft x 3ft 20swg mild steel sheets and set to work.

To get the bodyshell into the garage I used the ancient Egyptian log rolling method because it had neither wheels nor suspension. Inside, axle stands provided a comfortable working height and a steady, safe method of support. Closer inspection revealed rust and holes at all four corners, along the lower side seams and at each end of the box sections. There was not much left of the floor in the battery compartment and the forward rear subframe mounting points and the rear front subframe mounting points were severely weakened. To cap it all the engine stabiliser bar bracket had broken off. Was this, after all, a lost cause I asked myself? No, I needed the welding practise so what better way to gain experience than embarking on such a restoration or, should I say, ressurection?

Getting to work

The side seams were repaired first. They were tackled in 18in sections to preserve bodyshell strength and alignment. For cutting and grinding metal, especially in awkward places, I found an angle grinder essential. But for places that were relatively accessible a 7in grinding disc (with 5in backing disc) mounted on an electric drill was quite good enough. I always made a point of wearing goggles and ear protectors while doing this type of work

To save using a hammer and cold chisel I tackled the spot welded seams with an excellent Sykes-Pickavant spot weld removing tool costing about £5. The 18in sections of side seam were cut back to good metal which turned out to be about 1-2in from the edge. All the repair panels were cut with a power jigsaw and the top panel edge was set with a joddling tool to mate with the existing side panel.

All the welding was done with my trusty arc welder. I had to be particularly careful working on these thin panels to avoid burning right through. Another gadget I found useful was a vibrating welding electrode holder made by Keller. This pulses the arc, thereby reducing the risk of burning holes. It takes a lot of getting used to but the end results are well worth it.

Completing one side seam brought me to the front of the floorpan and the rear mounts of the front subframe. These mounting points are a water trap and, despite the drain holes, they rot. The easiest way to deal with this situation is to fold up a single piece of steel and box-in the whole mounting point. The bolt and drain holes can then be transferred to the new metal by drilling from the reverse side.

I tackled the floor rot next. The battery compartment was the worst but all areas called for straightforward patch repairs. However, the battery compartment also needed a new surround complete with nuts to take the panel fixing bolts. To obtain a neat folded edge I clamped the prepared replacement surround panels in a bench vice between two lengths of angle-iron and gently tapped them over with a heavy copper-faced hammer on to a block of wood. Longer panel folds called for something more specialised.

Proprietary folding machines are available of course but they are extremely expensive, extremely large and very, very heavy. I had access to one of these machines at work but generally I would advise that the folding is best done on site. This way the fit of the panel can be constantly checked on the car. So I made a small vice-mounted folder capable of producing a 2ft long bend using lengths of angle-iron and this proved quite adequate for the job.

After the other side seam had been replaced attention turned to the rear subframe front mounting points. The standard Mini repair panel for this was used to great effect. It is larger than necessary for the Moke but this is no bad thing as it allows plenty of tolerance. It was tack-welded into place first and the position was carefully checked before it was finally welded securely.

While the Moke was upside down I took the opportunity of welding the broken engine stabiliser bracket back on and thoroughly inspecting the underside. The underbody sealant was in a poor state and had to be removed. I used a hot air electric paint stripper for this which didn't take long and, with it all cleaned off, I applied two coats of the excellent Finnigan's No.1 rust inhibitor/primer. This was followed by two coats of Finnigan's underbody sealant which is a sort of very thick Waxoyl-type material that is applied by brush.

With the underside completed the body was righted and the other side of the engine stabiliser bracket could then be welded. The nearside rear wing showed tell-tale signs of being filled and, indeed, one heavy blow with a hammer revealed a foot-long dent which had been filled to the brim. This called for surgery. The affected area was cut out and an insert repair welded in, again setting the panel edges with the 'joddler'. I used my folding tool to form a flange with a returned edge so that it matched the rest of the wing edge.

Welding clamps are a must in this situation but I still found my longest clamp was lacking in reach. I had an old pair of Mole grips hanging around so, with the aid of some odd steel tube and the welder, I converted them into welding clamps with a 12in reach! This was enough to straddle the width of the wing and I haven't stopped using them since. They even came in handy when overhauling the gearbox, helping with bearing extractions of all things.

Where the front wings sweep down to the side boxes there was evidence of corrosion. So my next job was to cut this out and fit two triangular repair panels. Every repair of this type was followed with two application of, yes, you've guessed it, Finnigan's No. 1.

The next step was to strip the body of paint. Out came my hot air paint stripper again and it did the job well. A couple of swift passes with the scraper saw both yellow and the original green layers removed and, with a little more elbow grease, even the original primer was lifted away. The whole surface was then tidied with a flap

Making the Moke of it!

wheel/wire brush and finished with two coats of Finnigan's No.1 which was carefully applied by brush. Luckily this primer is very forgiving when it's brushed on and once it dried the end result was nice and flat in both senses of the word.

Panel making

The manufacture of the missing front body panels was the next step. To ensure all would be correctly aligned, the front subframe was hoisted up under the body and bolted into position using four out of the six mounting points. With the bonnet pressed on to the hinges the overall length of the panels could be judged. I had gathered some photographs of Mokes to determine dimensions and details and one of these was from the front cover of a "Blue Peter" annual of all things.

Patterns for the inner wings were cut from stout cardboard and placed into position for final shaping and trimming. Using the resulting template the shape was transferred to sheet steel and cut out with the jigsaw. Most of the folds were put in with my angle-iron metal bending tool. A radiused fold around the suspension tower area was achieved with careful application of a copper-faced hammer and a large diameter steel tube mounted in a bench vice.

I got the inner wings into position and was then able to make the measurements for the outer wings. Again, cardboard was used to obtain an accurate template. Once it was cut out the wings were bent into shape, remembering to fashion one for each side and not two the same! The wings were offered up to the inner wings and held in place with grip pins normally used when riveting aircraft structures.

Male and female wooden formers were fashioned to produce the seven louvres in the left-hand wing. Their positions were marked but, due to an error, eight slots were cut! With the female former on the outside of the panel the male half was placed inside and clouted with a heavy mallet. This procedure was repeated with each slot in turn, resulting in a neat row of perfectly-formed louvres.

A length of 1in square section tube was bolted to the front mounts of the subframe. Measurements were taken and compared with photographs to determine the grille shape and size and the positioning of the headlight and indicator units. All this information was transferred to steel and double checked before cutting began. The blank front panel was trial fitted – it was perfect. It was then removed and the headlight, grille and indicator holes were cut. To form a recessed grille aperture the edge was hammered into a wooden former using a blunt cold chisel. This added strength to the whole panel and provided a good location for the removable grille. The lower edge of the front panel was folded rearwards and the sides turned back with a returned flange. However, the top edge was left plain and the whole panel was then hung in position with grip pins.

The inner wings were back-marked for the shock absorber brackets and then removed, drilled and re-fitted. With the shock absorber brackets in place the front panels were checked for squareness with diagonal measurements taken across the engine compartment and all seams were welded up. A 1in wide strip of steel formed the top edge of the front panel to fill the gap in front of the bonnet. To this another strip was added thereby forming a flange for a bonnet seal. All the panel seams in the engine compartment were reinforced with $\frac{1}{2}$in x $\frac{1}{8}$in steel strip welded from the bulkhead, forward to the front panel then down to the subframe.

With the body once again upright a radiator cowl template was fashioned from cardboard. A steel copy was cut, bent and welded into position on the nearside inner wing. The external panel welds were ground down and smoothed with a finger wipe of body filler. However, the front panel suffered some heat distortion from the welding but this was corrected with a skim of filler and a couple of coats of Finnigan's No.1. The front end of the Moke was now complete and things were at last taking shape. It was a real morale booster!

The panels pinned into position while checking for accuracy of fit before welding.

Upside down is the perfect way to tackle the rot – the advantage of having a light bodyshell.

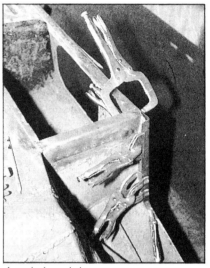
A typical repair in progress on a rotten corner.

By this time it was summer 1987 and the weather was ideal for spraying. Several coats of grey cellulose primer were followed by several more of Ford signal yellow – not the original colour I know, but I liked it! After letting it dry for a couple of weeks it was rubbed down with fine grade wet and dry and treated to five more coats. The bonnet and side panels were sprayed separately and the whole exercise represented my first attempt at spraying. I was more than satisfied with the results.

Back to the mechanicals

At the back a replacement rear subframe was painted with silver Hammerite and new brake pipes and hoses were fitted. The trailing arms received new pivot pins and bearings, and the seals in the brake cylinders and pressure regulating valve were replaced together with the handbrake cables. The front subframe and suspension were cleaned and reconditioned as were the brake components. The subframe-mounted sump guard was removed and given two coats of smooth yellow Hammerite and re-fitted and then the completed assembly was put to one side ready to accept the transmission.

Right: Fitting the hood top and back before making the sides and window apertures.

Below: The proud owner and his creation!

The plan was to use the original Moke transmission but, during the strip, something unexpected cropped up, or rather, poured out! As I gently tapped off the end cover from the engine I was greeted by a torrent of foul-smelling water instead of oil. A couple of gallons at least came out but, fearing the worst, I carried on and separated the engine and gearbox.

Every component inside was completely rusted and the whole thing was very depressing. However, all was not lost. I still had the transmission from the 'donor' Mini which I knew was a runner. I had hoped to use the Moke differential as it has the correct ratio but fate decreed I had to use the one from the Mini which was lower. So the 'Thompson Moke' would have neck-snapping acceleration but not a lot of top speed. But, then again, the Moke isn't a sports car is it?

The Mini's stripped engine was entrusted to a local engineering firm for reboring and crank grinding. Then, with the aid of an overhaul kit, I reconditioned the gearbox. It turned out to be an easier job than I had expected and the only special tools required were a two-legged puller and a torque wrench. I had to drill and tap holes in the transmission casing to accept the Moke sump guard and, using the guard as a template, these were drilled, paying particular attention to their depth so as not to drill right through.

Attention then turned to the freshly cleaned cylinder head and the valve guides were checked for wear. The valves themselves were lapped in and new stem seals fitted. By this time the engine block and crank had been returned complete with new shells, pistons and thrust washers and the whole thing was assembled. The engine and gearbox were mated, the flywheel and clutch fitted and the complete unit was lowered into the waiting subframe together with new transmission couplings.

The next step was to lift the body up and over the engine and back into its rightful place. It was light enough to be man-handled by me alone but it still needed considerable care to make sure all was well before bolting it down.

Brake and clutch master cylinders from the donor Mini were overhauled before being fitted along with the pedals and steering column support. Unfortunately, the original wiring loom had been cut just behind the bulkhead but I had already removed the required amount of loom from the donor, being careful to label every terminal. This multi-coloured spaghetti was laid roughly in position around the engine and through the bulkhead which allowed me to determine positions for the voltage regulator, starter solenoid, fusebox etc. The overall fit of the loom was quite good and the only modification needed was the lengthening of the wires to the indicators/sidelights and brake pressure switch.

The front and rear light units were bought from an autojumble for £4.50 and, having fitted the rear lights, I followed the wiring diagram colour coding to connect them up. Where the original loom had been cut I crimped it to the front loom according to the colours. Only one spare wire remained on the front loom and this was the interior light supply so it would now be redundant.

Interior fittings

The floor, inner sides and seats received the black Hammerite treatment to provide a hardwearing, durable surface. For a dashboard I decided to use a length of wood grain-patterned sheet steel taken from the side of an electric convector heater. In addition a metal parcel shelf was made and fitted which also doubled-up as a steering column and heater support. The Mini's demisting vents were positioned just below the windscreen and concealed behind more wood grain steel panelling. A box for the instruments was cut and formed from the remaining patterned steel and completed with slots on top to match the rest of the panels.

Upon installing the wiper motor and mechanism I found that the left-hand wiper spindle fouled the rear edge of the bonnet. It seemed that the latter was about a quarter of an inch too high. Indeed, when I checked this against other cars they seemed to have their bonnet hinge points mounted lower down on the scuttle than mine. Not being able to lower the hinge mounts I cut and lengthened the hinges on the bonnet which lowered the rear edge and provided the necessary clearance. Unfortunately this action also moved the bonnet forward slightly so it now rests on the front panel rather than flush with it as before.

At about this time by chance I came across a hefty coat rack. It was made from thick-walled steel tube and had four 90 degree bends and provided the ideal basis of a roll bar. The arc welder was pressed into service again and the required bends were cut from the rack along with the relevant straight sections to make the main 'goal posts'. It had to stand 3ft high and the ends were ground at an angle to produce a 'V' allowing for a good strong weld. A short length of angle-iron was clamped to the side of the joints to support it as each side was welded up. A rectangular mounting plate of ⅛in steel was welded to the base of each side tube and the whole assembly stood on the side box sections of the car just in front of the original hoop supports.

Two more tubes were manufactured with a 45 degree bend at one end and a mounting plate at the other. These were fitted running from the top of the 'goal posts' to the tops of the rear wings so providing fore and aft support. Two coats of black Hammerite made everything look neat and tidy and when it was finally bolted on some silicone sealant provided a watertight seal around the mounting plates.

A local glass supplier used the original windscreen glass as a pattern to cut a new laminated 'screen' for £24. I installed this and then set about making a set of quarter-light panels complete with Perspex windows.

I fitted inertia reel seatbelts up front and static ones in the rear. A brown camouflage material was used to make four padded seat covers all with dark brown edge piping. A friend gave me a brown 1275GT carpet set which, with a little trimming, was a satisfactory fit and a Lancia gearlever gaiter, held in place with a TR7 gaiter ring, tidied up this area.

The only item missing from the Moke now was a soft-top, the original being hopelessly rotten. Having access to an industrial sewing machine swayed me into making my own. Woolies of Market Deeping supplied all the vinyl, window material and lift dot fasteners for £165. I am glad that I made the effort to do it myself because it saved a lot of money and turned out very well. What's more, it doesn't leak!

I got the Moke insured and booked it in for an MoT check. However, it overheated on the way due to the wrong radiator cap being fitted and also the battery wasn't charging. Consequently, it had to be towed home where I sorted out the problems in a couple of days after which it made it to the MoT station and passed. The original registration number, OYU 525F, was not listed on the Swansea computer and so I was issued with CVG 98F instead which I am quite pleased with.

Overall the restoration took me about 18 months and cost approximately £770 and the Moke has been in use for the last two summers and great fun it is too!

DIY REBUILD

DONKEY WORK
ON A MINI MOKE

For years Mark Peacock had wanted a Mini Moke. David Hill discovered what happened when he finally got one.

There is always a reason for an individual's choice of classic, even though the individual concerned may not be aware of it. For example, the subliminal influence of Roger Moore's portrayal of Simon Templar probably lies behind the acquisition of many a Volvo P1800.

Perhaps the reason behind the choice may have been still more obscure; how many favoured yet long-lost Dinky toys have re-emerged in later years as the real thing? On the other hand, some budding car enthusiasts make a conscious decision at an early age as did Mark Peacock, with a little help from his barber. His barber? Yes, that had me wondering too. As it happened, Mark's coiffeur had a picture of a Moke on his wall. This illustration, apart from helping to diminish by distraction the injustice felt by a small boy forced to submit to the indignity of a severe haircut, also captured the young Mark's imagination. The seed was sown, to blossom in the fullness of time as a strong desire to own a Moke.

The transition from dream to reality took some time to occur, Mark having reached the ripe old age of 36 before he began a serious attempt to track down a suitable example. The pursuit initially took him to Crewe, to view two potential purchases. The Mokes involved lay at opposite ends of the buyer's scale: one was good but too expensive, the other was cheap but too rotten. Mark returned to Bradford temporarily worsted but nevertheless undeterred. Soon afterwards, he crossed the Pennines once more, to view a 1964 Moke in Wales. The car in question was closely examined and deemed to be suitable: having struck lucky on the third attempt, Mark returned to the land of song, armed with £650 and a flatbed trailer and before July 1988 was much older, the bargain was sealed.

Mark's prize boasted one month's road tax and an equivalent period of remaining MoT. A short exploratory trundle showed that it also had nothing to speak of in the way of braking so Mark made the wise decision to take the Moke off the road and restore it. This was not as painful a decision as might be thought because it was part of a master plan; Mark's intention was to carry out a complete restoration rather than just buy a Moke and use it.

When not in, around or under his Moke Mark Peacock is a colour matcher in textiles. By the time he'd attacked the Moke's metalwork, Mark saw quite a few more colours. Having hooked out the twin-carbed 1275c.c motor complete with gearbox, subframe and

● Mark kicks up a little dust whilst proving that a Moke is ridden rather than driven.

● The original 850cc A-Series motor was replaced with this 1275cc, A+ unit from an MG Metro. The standard head steady was supplemented with a hefty additional brace, fitted using the thermostat housing studs. The motor performs as well as it looks.

● Mark's Moke snoozes whilst its underside is assessed. The debris on the floor represents the tip of a sizeable iceberg.

● The tail panel had seen better days. Fortunately, the area complicated by vertical pressed flutes could be saved.

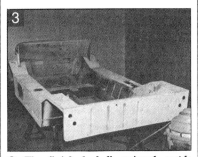

● The finished shell, primed, guide coated and about to be sprayed and oven-baked in two-pack Flame Red.

● Wooden pallets were used to bring the Moke up to a convenient working height. The aperture in the pontoon had to be specially cut in a later-spec. panel.

all the whizzy bits, Mark began removing underseal. This revealed that the car had been finished in yellow, over red...over blue...over green. Eventually, he struck primer and bare metal, except where he struck rot or fresh air.

Needless to say the car was ripe for restoration, having ripened rather too extensively for its own good in the rear subframe area and floorpan. The notorious tinworm had also been at work on the rear panel and boot floor and had followed its hors d'oeuvre by making a meal of the floorpan before having the inner wings for dessert. That was the bad news. The good news exists in the Moke's simplicity; for the majority of the panelwork, the description 'flat' is entirely appropriate.

Amateur boxing

Mark began the rectification process by removing the floorpan, leaving only the internal centre crossmember. This provided a suitable reference point for fitting of a quartet of new floor pans and new transmission tunnel Mark had had made locally. A proprietary Mini seat pan front panel was used to equip the Moke with new mounts for the forward end of its rear subframe and Mark used the remains of the original rear subframe as a jig to align his new, genuine rear floor panel. The establishment of the new rear floor in the correct place allowed Mark to accurately place the rear subframe mounts and the repair work continued as far as the bottom edge of the rear panel. So far, so good.

The rigidity of the Moke's floorpan is only partially provided by the tunnel pressing. The lion's share of the task goes to the side sections, which could be considered as overgrown sills but are probably best referred to as pontoons. These were found to be holed, so the next exercise involved replacing their lower extremities. Next, Mark addressed himself to the rear mounts of the front subframe, which had suffered as only double skins can. The minimal remains of the original mounts were hopeless as patterns so Mark, having resorted to examining another Moke for reference, used the front subframe as a jig to align his new mounts and repaired the inner wings by way of an encore.

The work to date had replaced the strength of the main structure. In fact, it was only then that Mark considered the shell stiff enough to be blast cleaned. The blasting revealed rot in the outer sides of the pontoons - which Mark had already spotted - and on the tops, where the wings join the pontoons at either end. These problems came as a surprise but were cured with a spot of localised repair work. Ron Smith at Runamoke, who had supplied the boot floor panel, sent a pair of new pontoon sides, which were as per the later, Portuguese-built, Moke, lacking the hatch in the forward pressing which allowed the fuel pump to live in the nearside pontoon. The nearside panel also had provision for a side-entry fuel filler - unnecessary on Mark's car which had a filler atop the nearside pontoon. So, the appropriate modification consisted of cutting one hole and blanking off the other but the fitting of the side panels was postponed for the present. Mark finished off the steelwork by making a new lower edge for the front panel. This was stepped with a joddler and fitted flush with the original skin. Lastly, Mark replaced the rail which accepts the lower edge of the screen.

Until now, Mark had used a 120 amp MIG welder and CO2 for shielding. While he could have plug-welded the new pontoon sides, he felt that spot welds would be neater in such a highly visible area. With this in mind, the car was taken to nearby Hatfield Motors. Here, the sides were spot-welded on before the shell, on which Mark had already sealed the seams, was etch primed and painted. The chosen colour was BL Flame Red in two-pack, which was applied after the shell had been protected with stone-chip resisting paint, both inside and underneath.

During its primed period, the Moke had been resting on axle stands and bits of wood - good enough for a shell being positioned variously on its side and upside down. In deference to the new paint, Mark borrowed three clean pallets to place the bare shell at a convenient height.

Since buying the Moke, Mark had also acquired a number of Mini Cooper parts which were sold at Cooper prices. This funded the purchase of a rebuilt MG Metro motor, along with a new starter and alternator. Mark's colour-matching skills came in handy at this point, to mix Humbrol enamels to the appropriate MG spec for painting the motor: For the record, the mix included red, orange and yellow. In fitting the engine, Mark encountered a small but significant problem. The new tunnel had been beautifully fabricated-with no gearlever aperture. Careful measurement and cutting solved the problem.

Meanwhile another local firm, Vernon Moss, were busy chroming various bits, including the bumpers and steering column. Another local concern had been equally busy. Mark had been very careful with the screen glass, covering it up with a blanket to avoid damage from flying sparks in the workshop.

DONKEY WORK
ON A MINI MOKE

Despite this precaution, the screen had suffered a bad case of grinder rash so the firm in question used it as a pattern to cut a new screen from laminated glass. The proprietor, whilst laying the new screen on the old to show that the cutting was accurate, inadvertently put pressure on the glass and...crack!

The company had no more clear laminated glass in stock so you now know why Mark's Moke has a tinted screen. A second accident with the screen became a real possibility during the fitment. The process involved fitting the well-lubricated screen rubber to the glass, taking it off and fitting it to the frame, then back to the glass. This method failed, so Mark resorted to sitting on the frame in the company of a good friend to compress the rubber, allowing the bolts to be nipped up before anything broke.

By now, the Moke had an engine, front and rear suspension with all-new bushes and bearings, and an exhaust. Before long, it had a braking system, a wiring loom rescued from the local breaker's, new headlamps in rot-proof plastic mounts and new rearlights. The original, steel front seat shells had been replaced, accompanied by fibreglass copies for the rear, and the front sidelight backings had also been replaced. All-new rearlights were fitted and Mark's newly-purchased Land Rover wiper mechanism and fuel filler cap had returned from the plater's. Soon, the Moke was travelling the length of Mark's drive under its own power, a milestone in the restoration.

The need for something less punishing than bare steel to sit on became pressing and prompted the involvement of a friend's sister who obligingly stitched together a set of seat pads, completing the task to a high standard despite being an absolute beginner at car trimming. The hood was farmed out to another local outfit, Autotrim in Bradford (Tel: 0274 480106). The result was a remarkable achievement in that it fitted perfectly, first time. Why remarkable? Because the new hood was made using only the sagging, threadbare original as a pattern, without the benefit of the car's presence - nice one, Autotrim.

By the time last June blazed its torpid way into the meteorological record books, the Moke was up and running, taxed and tested. On the day I saw it, the car had all of twenty miles on its legally-clocked odometer and was ready to pose. The posing took place at a nearby reservoir, by courtesy of the local sailing club who kindly permitted us to use their causeway as a location. On the way, the Moke, carrying Mark and his friend and fellow Moketeer, Geoff Priestley, showed me a clean pair of wheels, tight engine notwithstanding. Incidentally, talking of wheels, the Moke normally wears a set of Mamba alloys. In the cause of originality, Mark had fitted his newly-refinished 10in wheels for the pictures.

Later, Mark took me over the same roads on which he'd originally tested the Moke, four years ago. During this trip the Moke displaying a well-controlled ride, sharp handling and no untoward behaviour. The 1275cc lump, although by no means original, echoes the specification of the Australian-built Mokes and their successors. More to the point, it is quieter, more powerful and more refined than the original 850 cc A-Series engine. Having driven an original Moke, I found those differences striking and the photographs show the high standard achieved. Others clearly felt the same way about Mark's work; his Moke has already won two first prizes having been considered the best English Moke, at a recent Mini Owners Club event and at the National Mini Moke Weekend at Ironbridge in Shropshire.

Mark carried out his restoration during evenings, weekends and holidays, and he and Geoff found the exercise enjoyable, citing the removal of the underseal as the least attractive aspect. Like many full restorations, it's not quite over - Mark is still looking for the correct, green-topped indicator stalk and wants an original steering wheel. Costwise, the restoration had been gratifyingly restrained. Mark's outlay multiplied the initial buying price by a little over four, giving a total of around £2,800. The results are visible enough; what you can't see is the Moke's best aspect, i.e. it's good fun, both to own and drive. Moreover, lacking doors, roof, A,B and C, posts, not to mention curves, it's easy to restore and can be recommended as a good entry-level project.

Finally, Mark would like to use this opportunity to thank Geoff, for his help and support during the restoration and his neighbour, John, for the extended loan of the MIG welder. Mark is also indebted to Autotrim, Frank Davis at Hatfield Motors and fellow members of the Mini Moke Club. And last but certainly not least, Mark's thanks go to his wife, Elaine, for putting up with the fulfilment of a boyhood dream. ∎

Clockwise from the top:

● *Not much work for a trimmer here. Mark shortened the gearchange rods as per later, Australian-built Moke spec. Land-Rover wiper mechanisms (and filler caps) were standard equipment.*

● *Mark and Geoff demonstrate the Moke's fast-action hood, which can be stowed in a matter of seconds. When it comes to owning such a car, only fresh air fiends need apply.*

● *The final result, complete with hood and original road wheels. Mark has no intention of fitting side curtains.*

THESE AUGUST PAGES have, from time to time, featured analyses of particular marque ownership. They usually portray a typical owner and his joys and disappointments as he participates over some extended time in this delightful sport of ours. The articles typically yield some precise data in terms of cost per mile of ownership, and they give another fellow some idea of what he might expect in owning a similar car. It happens that neither am I a typical fellow nor is the vehicle I treat here particularly commonplace. Yet my analysis might well change the entire rationale of automobile purchasing.

First off, I live and motor on a tiny dot in the Caribbean. The island I currently call home is roughly the size of Manhattan, and its topography is suggested well by imagining a gigantic sheet draped over the latter, skyscrapers and all. Our roads are narrow, hilly and bumpy, and to add to the general excitement of motorsports we are U.S. citizens who use left-hand drive on the wrong side of the road. There would be little chance of a 50,000-mile extended test originating here, for no one has ever had a car hold up for that kind of mileage. We do, however, have a spirited array of motorsport activities, ranging from Downhill-Driveway Trials (like bob-sledding, brake use means disqualification) to the awesome Shugly-Mon-Up 'n'.Down Hillclimb. The event derives its name from local dialect: "Dat road, she ugly, mon, up 'n' down!" But I stray from the point.

Our automotive tastes are largely dictated by the road conditions. For example, none of the local Healeys, sample size of two, has an exhaust system beyond the manifold. My '67 Volvo wagon is holding up rather well, but it will never see that eleventh birthday party the ads promise. VWs abound; it seems that their numbers are always directly proportional to the primitivity of road conditions. Another suitable vehicle, (there are those who balk at calling it a car) is the BMC Mini-Moke. For those readers unfamiliar with the marque, imagine a runt offspring of Jeep out of Mini, and you're quite close. Its open-air qualities suit our climate, its ground clearance our driveways.

One never buys a Moke new. Because of the marque's endearing mechanical charms, there are always scads of us scrambling to move up to a VW, which pretty well sums up both Mokes and us. My acquisition was typical: we wanted a second vehicle, but our limited need precluded buying a car. A motorcycle would have sufficed, but that was before I summoned the courage to try one on our roads. Hence a Moke was the

JON DAHLSTROM ILLUSTRATION

After the New Gets Put On Again and Wears Off For Good: MINI-MOKE

BY D.J. SIMS

perfect choice, and there were several for sale. A quick leap to 65 along the golf-course road narrowed the choice, as well as the road, and for $200 it was mine. The leap was especially exciting because Mini, as she came to be known, had neither windscreen nor muffler. She did have a certain pre-vintage charm, though. I can see why people bought those tatty little cyclecars during the Twenties: there is genuine satisfaction, not to say occasional excitement, in urging a pint to do a quart's work.

Like a Damon Runyon character's horse tip, all Mokes come with a story, and mine was no exception. She was brought on-island by a rental agency; to the best of my research, all our Mokes were. After having been thrashed by carefree vacationers for 8000 miles she was sold to a student of mine who promptly stripped her for the Shugly-Mon. He refinished her in the colors of his equipe, Dayglo orange and flat black (he had been born on Halloween), and in her debut he placed second to a Rickman-Matisse Scrambler motorcycle. If nothing else, our entry lists were catholic!

After 700 miles of activities, surely no worse on her than a couple Mille Migliás, she passed into my possession. My student friend had tired of the Caribbean and transferred to a college in Reykjavik, Iceland.

My tastes are basically of a vintage turn, and I immediately planned a restoration of Mini's road trim. This proved complicated, for parts, like complete Mokes, were rarely obtained new. In fact, the general practice of stripping abandoned cars rivals drinking rum as the local sport. Such random acquisition goes against my nature, though. After all, the other guy abandoned the thing for reasons other than material renunciation; I don't need *his* troubles. Eventually I was able to buy, and in a few cases let us say acquire, the parts needed for Mini's restoration. The work began with a paint job of BRG with cream trim; I have this thing about a Morgan in just such a scheme. Like the Mog of my dreams, Mini was painted with a brush, actually two of them. She turned out rather less finished than a Malvern product, but a damn sight better than her Equipe Goblin colors. Mechanically, she was in fair shape except for ominous sounds in the gearbox. (They always seem to go snick in R&T; mine went ger-reek-ching.) I traced the problem to weak hydraulics in the clutch and solved it by learning to change up and down without this troublesome contraption. My total cost of restoration, including parts, paint and brushes, amounted to $64.13. For the first time since her rental days, Mini underwent a legal inspection at 8721 miles. I celebrated the occasion by sitting in her for two hours afterward while I enjoyed the world's largest martini. The neighbors had grown used to my eccentricities; they pretended not to notice.

Mini provided charming pre-vintage motoring for a while. Her weakest area seemed to be in braking, or rather in the delayed application of this activity. Through some quirk of age and hydraulics, it was not uncommon for the brakes to apply themselves an instant *after* the pedal was depressed. As one vintage writer so aptly put it, this demanded "a certain precognition of traffic patterns," rather like the opposite of driving a turbine with throttle lag. Apart from this idiosyncrasy, Mini gave me dependable transportation for some 103 miles.

During this time she and I took part in some informal rock climbs but avoided running the Shugly-Mon which had become an infrequent event held late at night. I began to notice some front-end scrubbing and at 8824 miles took a good head-on look at her. One wheel pointed out at an alarming angle but the other looked, more or less, toed-in. Evidently the rock climbs had played havoc with alignment, and I proceeded to adjust a nice even toe-in to both wheels. (The work cost $1.25; I had to buy a decent tape measure.) It left things worse than before, and I gave up serious maintenance for a weekend of relaxation. Some of the time was spent leafing through motor magazines, and an article about front-end geometry caught my eye. I read all about the theory of caster, camber, contact patch and toe-in/toe-out. Toe-out?? Sure, fwd cars need a bit so that the contact patch will pull the wheels parallel under power. Fwd??

Mini is fwd! Five minutes work had her tracking properly again.

Then came the vintage phase of my ownership. She ran superbly for almost 1000 miles, and I gained much respect from my peers, for she was the only Moke we knew that had ever been waxed (9438 miles, $1.38, wax).

But at the reading of 9624 there was a quick sequence of mechanical calamities which necessitated a tidy investment; this time she required *new* parts. My wife and I had visited a beach at the end of a steep rocky path, and while we were there one of our island's passing downpours occurred. They drop gobs of water in 30 seconds, then drift elsewhere. This one had left the return road in rather worse shape than when we had descended. Our rock climb experience came to good use (racing improves the breed and all that) and we proceeded to storm up backward with my wife leaning out over the hood. Indeed, this was ordinary rock technique for a Moke, what with fwd, weight transfer and its absurd weight distribution in the first place. We got about halfway up when my wife decided, in a rather general sort of way, to walk.

Her falling out caused the motor to roar, in one of those complicated sequences of mechanical events I call the King Richard III Syndrome. You may recall that eventual outcome when a mount loses a nail, and losing my wife at that instant had similar consequences. The sudden weight loss caused great bouncing wheelies of the front end. This in turn put undue strain on all those damned drivetrain doughnuts of which the British are so fond. One of the doughnuts yielded under the strain and the resulting shock to the engine snapped its top mount. As it twisted on the two bottom mounts, it broke its connection with the muffler, and hence the roar. We got home eventually, but it wasn't all the noise which precluded conversation. It took the tidy little sum of $27.52 to purchase new doughnuts and to have fabricated an engine mount which could do justice to a 747. Neither caused any further trouble, although I should add that the last 400 miles of motoring were done monoposto.

It was this solo motoring which led to Mini's leaving our family. We had finally moved up to a VW, and simultaneously a friend gifted me with a collection of boxes allegedly containing a Honda Scrambler. (By this time I was over my rock-riding fears, and another restoration was in the planning.) Our equipe now consisted of a Volvo wagon, a VW, the Honda in the boxes, and Mini. Clearly one too many, and only Mini seemed expendable.

At 10,143 miles, I sold her for $350 to a colleague who was known for his fatalistic view of things mechanical (he bought an open-end wrench once; it broke). He had relatively trouble-free use of Mini for a while and then moved up to a Renault which, more typically for him, threw a rod. Lately, I've seen Mini being driven by an interesting-looking gentleman who affects a George Washington hair style.

My experience of ownership has interesting implications, as the table indicates. It seems that Mini cost me a total of −$32.42, which says, in effect, that I was being paid about 2¢ per mile for driving her. In retrospect, that seems just about right.

MINI-MOKE
Overall Costs & Cost per Mile for 1428 Miles

Initial price (8715 miles)	$200.00
Restoration (8721)	64.13
Front-end alignment, tape measure (8824)	1.25
Wax (9438)	1.38
Two doughnuts (9624)	23.52
Engine support (9624)	4.00
Gas, oil, license	23.30
Total expenditure, 1428 miles	$317.58
Resale value, 10,143 miles	$350.00
Cost of driving 1428 miles	−$32.42
Overall cost per mile for 1428 miles	−2.27¢

ABOUT BROOKLANDS BOOKS

As a company Brooklands Books are dedicated to preserving motoring literature for enthusiasts. They have in print over 700 titles which deal with various aspects of motoring. A guide to each series is given below.

REGULAR BROOKLANDS BOOKS. Collected road tests, model descriptions, driving impressions and other articles on individual models and marques - plus muscle car compilations - taken from a wide variety of car magazines in Britain, North America, South Africa and Australasia. Filled with informed and reliable comment, specifications, performance data, buying used reports and other details. These books provide a fine opportunity to absorb all relevant information on the car you own, are thinking of buying or restoring, or perhaps simply admire.

BROOKLANDS 'COLLECTIONS'. Similar to 'Regular' Brooklands books, but lower priced and fewer pages.

GOLD PORTFOLIO SERIES. Again drawing on material from a wide range of motoring magazines, books in this series are devoted to prestige and high-performance cars from Britain, the Continent and North America. They have almost twice as many pages and illustrations as 'Regular' Brooklands Books and are an unparalleled source of interest and knowledge on the world's finest and most sought-after cars.

MUSCLE, PERFORMANCE & MILITARY PORTFOLIOS. The titles in these categories deal respectively with US and British muscle cars, performance marques and military subjects and have half as many articles again as the 'Regular' Brooklands Books.

PRACTICAL CLASSICS SERIES. Comprehensive restorers' guides, compiled from the pages of 'Practical Classics' magazine and based on 'hands-on' experience. All restoration tasks are covered.

ROAD & TRACK SERIES. Collected tests, comparisons, new model introductions, racing and touring articles and other material from the internationally popular American magazine 'Road & Track'.

CAR & DRIVER SERIES. Similar collections from 'Car & Driver' magazine, USA. Once again, these books supply a wealth of information on numerous models, in a package easy both to read and to keep.

CYCLE WORLD SERIES. Compiled exclusively from the US magazine 'Cycle World'. Included: on and off-road tests, technical information, touring and racing articles, comparisons and other material on a wide range of bikes.

HOT ROD ON GREAT AMERICAN ENGINES SERIES. Musclecar and high performance material drawn from Hot Rod, Car Craft and other Petersen publications on how to maintain, tune, repair, restore and modify the popular power plants of the musclecar era.

BROOKLANDS TECHNICAL BOOKS. These are all originals or unedited reissues of factory publications, including workshop manuals, handbooks and parts catalogues. Far more detailed than the 'condensed' literature widely available, they are essential possessions for owners and restorers and are highly recommended. We also include some official sales brochures within this series.

OWNERS WORKSHOP MANUALS. Brooklands Books have acquired the rights to the Autobooks Workshop Manual series plus some manuals originally published by Chilton and Clymer. Written and illustrated by experts, they cover servicing, overhaul and repair for DIY enthusiasts. Many are now reissued in larger format with an all-new chapter on buying a used model or as handy 'glovebox' editions.

BROOKLANDS GENERAL RESTORATION & TECHNICAL BOOKS. Practical guides for owners wishing to restore or upgrade their cars. Most of these titles are drawn from the pages of 'Hot Rod' magazine and fall into our 'Restoration Tips & Techniques' or 'Performance Tuning' categories. All contain detailed and expert text, together with hundreds of explanatory photos.

BROOKLANDS REISSUES OF MAGAZINES. Early copies of 'Road & Track, 'Hot Rod' and 'Autosport'. Economically priced, these reissues are a boon to collectors.

Send for catalogue and full list to:

Brooklands Books Ltd., PO Box 146, Cobham, Surrey KT11 1LG England. Phone: 0932 865051. Fax: 0932 868803
Brooklands Books Ltd., 1/81 Darley Street, PO Box 199, Mona Vale, NSW 2103, Australia. Phone: 2 997 8428. Fax: 2 979 5799
CarTech, 11481 Kost Dam Road, North Branch, MN 55056, USA. Phone: 800 551 4754 & 612 583 3471 Fax: 612 583 2023
Motorbooks International P O Box 1, Osceola, WI 54020 Phone: 800 826 6600 & 715 294 3345 Fax: 715 294 4448

BROOKLANDS BOOKS

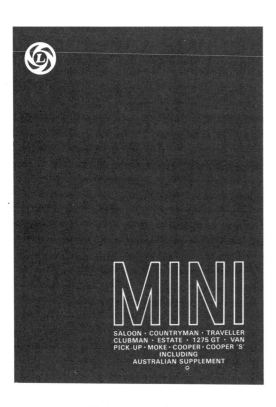

MINI OFFICIAL WORKSHOP MANUAL (1959-76) 9th edn. The complete professional or amateur mechanic's guide to all repair and sevicing procedures. Covers Saloon, Countryman, Traveller, Clubman, Estate, 1275GT, Van Pick-up, Moke, Cooper 'S'. 362 pages. SB. (AKD 4935)
Ref. B-M13WH

MINI OFFICIAL WORKSHOP MANUAL (INC. AUSTRALIAN SUPPLEMENT) As above, but with an additional 178 page Australian supplement. (AKD 4935/2) SB.
BEST BOOK FOR MOKE OWNERS
Ref. B-M13AWH

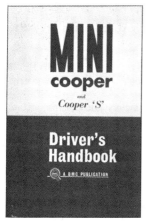

MINI COOPER & COOPER 'S' OFFICIAL OWNER'S HANDBOOK
Pub. 1967. routine servicing, data, 72 pages, illus. SB. (AKD 3891G)
Ref. B-M14HH

MINI COOPER 'S' MK.II OFFICIAL OWNER'S HANDBOOK
Controls, care, routine maintenance, data. 56 pages, drawings. SB .(AKD 4997)
Ref. B-M15HH

MINI COOPER 'S' MK.3 OFFICIAL OWNER'S HANDBOOK
Full information on care and maintenance, plus general data. 68 pages, drawings. SB. (AKD 7364 2nd edn.)
Ref. B-M16HH

From specialist booksellers, or in case of diffiuclty, direct from the distributors:
Brooklands Books Ltd., PO Box 146, Cobham, Surrey KT11 1LG, England Tel: 0932 865051
Brooklands Books Ltd., 1/81 Darley Street., Mona Vale, NSW 2103, Australia. Tel: 2 997 8428
CarTech, 11481 Kost Dam Road, North Branch, MN. 55056, USA. Tel: 800551 4754 & 612 583 3471

WORLDS LARGEST MOTORING REFERENCE LIBRARY